Süßwasserflora von Mitteleuropa

Bd./Vol. 1/1	Chrysophyte and Haptophyte Algae, 2nd ed., Part 1 (Chrysophyceae s. str., Dictyochophyceae, Phaeothamniophyceae and Haptophyceae)
Bd./Vol. 1/2	Chrysophyte and Haptophyte Algae, 2nd ed., Part 2 (Synurophyceae) (2007)
Bd./Vol. 2/1	Bacillariophyceae (Naviculaceae) (1999)
Bd./Vol. 2/2	Bacillariophyceae (Bacillariaceae, Epithemiaceae, Surirellaceae) (1999)
Bd./Vol. 2/3	Bacillariophyceae (Centrales; Fragilariaceae, Eunotiaceae) (2004)
Bd./Vol. 2/4	Bacillariophyceae (Achnanthaceae, Literaturverzeichnis) (2004)
Bd./Vol. 2/5	Bacillariophyceae (English and French translation of the keys) (2000)
Bd./Vol. 3	Xanthophyceae 1. Teil (1978)
Bd./Vol. 4	Xanthophyceae 2. Teil (1980)
Bd./Vol. 5	Cryptophyceae und Raphidophyeae
Bd./Vol. 6	Dinophyceae (Dinoflagellida) (1990)
Bd./Vol. 7	Rhodophyta and Phaeophyceae (2011)
Bd./Vol. 8	Euglenophyceae
Bd./Vol. 9	Chlorophyta I (Phytomonadina) (1983)
Bd./Vol. 10	Chlorophyta II (Tetrasporales, Chlorococcales, Gloeodendrales) (1988)
Bd./Vol. 11	Chlorophyta III (Chlorophyceae p. p.: Chlorellales, Protosiphonales)
Bd./Vol. 12	Chlorophyta IV (Chlorophyceae p. p.: Stichococcales, Microsporales; Codiolophyceae: Ulotrichales, Monostromatales)
Bd./Vol. 13	Chlorophyta V (Chlorophyceae p. p.: Chaetophorales, Trentepholiales, Chlorosphaerales)
Bd./Vol. 14	Chlorophyta VI (Oedogoniophyceae: Oedogoniales) (1985)
Bd./Vol. 15	Chlorophyta VII (Bryopsidophyceae: Cladophorales, Sphaeropleales)
Bd./Vol. 16	Chlorophyta VIII (Conjugatophyceae I: Zygnemales) (1984)
Bd./Vol. 17	Chlorophyta IX (Conjugatophyceae II: Zygnematales: Mesotaeniaceae; Desmidiales)
Bd./Vol. 18	Charales (Charophyceae) (1997)
Bd./Vol. 19/1	Cyanoprokaryota I (Chroococcales) (1998)
Bd./Vol. 19/2	Cyanoprokaryota II (Oscillatoriales) (2005)
Bd./Vol. 19/3	Cyanoprokaryota III (Nostocales, Stigonematales)
Bd./Vol. 20	Schizomycetes (1982)
Bd./Vol. 21/1	Fungi (Lichens)
Bd./Vol. 21/2	Fungi
Bd./Vol. 22	Bryophyta
Bd./Vol. 23	Pterido- und Anthophyta 1. Teil: Lycopodiaceae bis Orchidaceae (1980)
Bd./Vol. 24	Pterido- und Anthophyta 2. Teil: Saururaceae bis Asteraceae (1981)

Pertti Eloranta / Janina Kwandrans /
Elsalore Kusel-Fetzmann

Rhodophyta and Phaeophyceae

Süßwasserflora von Mitteleuropa

Freshwater Flora of Central Europe

Begründet von / Founded by A. Pascher

Herausgegeben von / Edited by
Burkhard Büdel · Georg Gärtner
Lothar Krienitz · Hans-Rudolf Preisig
Michael Schagerl

Band / Volume 7

Pertti Eloranta
Janina Kwandrans
Elsalore Kusel-Fetzmann

Rhodophyta and Phaeophyceae

Rhodophyta and Phaeophyceae

Pertti Eloranta
Janina Kwandrans
Elsalore Kusel-Fetzmann

346 Abbildungen

Spektrum
AKADEMISCHER VERLAG

Authors' Adresses
Professor em. Dr. **Pertti Eloranta**, Sinkilätie 13, 40530 Jyväskylä, Finland
Professor Dr. **Janina Kwandrans**, Institute of Nature Conservation,
Polish Academy of Sciences, al. A. Mickiewicza 33, 31-120 Kraków, Poland
Professor Mag. rer. nat. Dr. phil. **Elsalore Kusel-Fetzmann**, In den Messerern 9,
2340 Mödling, Austria

Volume Editor
a.o. Univ. Professor Mag. Dr. **Michael Schagerl**, Universität Wien, Althanstraße 14,
1090 Wien/Österreich

Bibliographic information published by Die Deutsche Nationalbibliothek
Die Deutsche Nationalbibliothek lists this publication in the Deutsche Nationalbibliografie;
Detailed bibliographic data is available in the internet at http://dnb.d-nb.de.

Springer is a part of Springer Science+Business Media
springer.de

11 12 13 14 15 5 4 3 2 1

Acquisition Editor / Developmental Editor: Dr. Ulrich G. Moltmann / Dr. Christoph Iven
Composed by: SatzReproService GmbH, Jena, Germany
Illustrations: by the authors (if not stated otherwise in the figure captions)
Cover Design: SpieszDesign, Neu-Ulm, Germany

ISBN 978-3-8274-2051-0

Editor's Preface

In today's world, in which scientists are evaluated based on Impact Factors and journal output, it has become increasingly difficult to devote efforts to monographs, the preparation of which requires time-consuming studies and decades-long experience.

Three outstanding scientists were recruited for the present volume. They combine expertise for the brown and red algae and accepted the challenge to prepare this monograph, situated at the crossroads between traditional and modern taxonomy. Such a publication can never be adequately remunerated, nor will it receive adequate acknowledgement in the latest evaluation schemes. It is, however, a work that will serve as a valuable resource worldwide and be consulted time and time again by many interested readers. This profound effort is destined to become a lasting contribution.

It was an honor and pleasure for me to be entrusted with editing this volume and to work as a team with Janina Kwandrans, Pertti Eloranta and Elsalore Kusel-Fetzmann – I thank you all!

Michael Schagerl
Responsible Editor

Preface to Part I, Rhodophyta

The research activities on the freshwater rhodophytes in Europe have been rather sparse during recent decades. The group is small and plays no major role in evaluating and monitoring water quality as compared with diatoms or cyanoprokaryotes. Nonetheless, several freshwater rhodophyte taxa are included on the lists of threatened species in many countries. Man-made changes in aquatic environments are the main reasons for the disappearance of many rhodophyte species, especially in Central Europe. Many taxa preferentially inhabit small rivers and creeks with unpolluted waters, which today are mainly found in sparsely populated mountain areas.

One reason for the scarce research on the freshwater red algae is the lack of suitable taxonomic literature. A case in point – found in floristic inventories – is 'common' names like *Batrachospermum moniliforme* and *Lemanea fluviatilis*: they often indicate the occurrence of the mentioned genera but lack the proper identification down to the species level. The previous edition of the 'Süsswasserflora'-series including rhodophytes dates back to 1925. Later, two more keys were published; one by Karol Starmach in the series of Polish Flora in 1977, and more recently another one by Shigeru Kumano in 2002. The numerous changes in the continuously developing taxonomy are evident in numerous synonymies and frequent species revisions. The traditional taxonomy based on morphometric characteristics has received new insights from gene sequencing, which also provides information about the phylogeny and relationships between the taxa.

The authors would like to express their warmest thanks to many algologists of European countries for interesting samples, permissions to use their photos and support to collate literature on freshwater red algae in their countries. Our special thanks go to Roland Bengtsson from Sweden, Maria Aboal and Sergei Sabater from Spain, Hans-Rudolf Preisig from Switzerland, Frantisek Hindák from Slovakia, Brian Whitton and David John from Great Britain, Luc Ector from Luxemburg, Keve T. Kiss from Hungary and Jolanta Kostkevičienė from Lithuania. The Władysław Szafer Institute of Botany, Polish Academy of Sciences in Kraków is acknowledged for permission to use figures by Karol Starmach (1977).

The authors are grateful to Ms Barbara Biernat for revising the English language in the introduction part and to Professor Dr. Michael Schagerl and to Dr. Michael Stachowitsch for polishing the English of the whole text. The final step of the book processing was carried out within the scientific project No. N N304 285937 supported by the Polish Ministry of Science and Higher Education.

Pertti Eloranta / Janina Kwandrans

Preface to Part II, Phaeophyceae

The predominantly marine class brown algae *(Phaeophyceae)* is placed within the phylum *Heterokontophyta* (or more clearly defined *Ochrophyta*, Graham et Wilcox 2000, or *photosynthetic Stramenopiles*, Graham et al. 2009) and consists of 265 genera comprising about 2000 species. Zoospores and gametes are the only unicellular forms and show active movement; the vegetative stages are always multicellular. Phaeophyceae cover a wide size range from tiny taxa, recognizable only under a microscope, to huge seaweeds (kelps) that reach many metres in length. So-called kelp forests can be found in certain temperate coastal zones associated with nutrient-rich currents. Commonly, Phaeophyceae grow attached to rocks or other hard substrates in the intertidal and subtidal coastal zones of temperate, boreal and polar regions (Womersley 1987).

From freshwater habitats, only 7 species (within 5 genera) have been described (Van den Hoek et al. 1995). These taxa are small, inconspicuous, with creeping or upright filaments, forming tiny cushions, tufts or flat crusts on stones. Some forms are endo- or epiphytic. Freshwater Phaeophyceae are easily overlooked, but may be common in fast-flowing rivers and along the rocky shores of lakes.

Since Pascher (1925a), Bourrelly (1981) and Starmach (1977), no monographic revision of European freshwater Phaeophyceae has been published; Wehr (2003) revised this group for North America.

The classification of brown algae is still based on morphological traits and life history, but molecular methods have already revealed a different picture of phylogenetic relationships compared to the traditional identification keys. Nonetheless, further work, especially for freshwater brown algae, is needed before phylogenetic relationships can be resolved (McCauley & Wehr 2007). This issue treats the species of freshwater brown algae, regardless of their final systematic position.

Finally I wish to thank all who were helpful to me. Dr. Marieluise Weidinger many thanks for the electron microscopical photographs. Prof. Michael Schagerl many thanks for the layout of the photograph and drawing composites; Mag. Manuela Zinöcker thanks for adjusting the lettering of the tables. Prof. Michael Schagerl further many thanks for patient and critical checking and correction of the English manuscript. Thanks also to Doz. Michael Stachowitsch for final English proof reading. Prof. Dieter G. Müller, Konstanz, many thanks for the present of a living piece of the Australien Ectocarpus. I am very grateful to Dr. Akira Peters for interesting and valuable discussion. To my husband Dr. Hermann Kusel I am very thankful for far away excursions together to take samples of freshwater-Phaeophyceae and for his patience with me at working with the algae.

Elsalore Kusel-Fetzmann

Contents

Part I Rhodophyta

Pertti Eloranta and Janina Kwandrans

Introduction

Studies on freshwater red algae, particularly in Europe

Linnaeus (1753) described fourteen genera of algae, and red algae were treated under the genera *Conferva, Fucus* and *Ulva*. Lamoroux (1813) segregated red algae due to their colour and morphology from the remaining algae and named this group Floridées. Sirodot (1884) paved the way to modern taxonomy of the family Batrachospermaceae and grouped the species of the genus *Batrachospermum* into six sections: *Moniliformes*, *Helminthoides*, *Setaces*, *Turficoles*, *Verts* and *Hybride*.
Pascher (1925) published a monograph on freshwater algae of Europe entitled "Die Süsswasser-Flora Deutschlands, Österreichs und der Schweiz", and freshwater red algae were considered together with Heterokontae, Phaeophyta and Charophyta (Pascher & Schiller 1925). In that book, freshwater red algae were divided into the two orders Bangiales and Floridae. Kylin studied rhodophytes, especially in Swedish fresh and coastal waters from 1906 to 1956. Skuja published a large number of taxonomic and floristic articles on freshwater red algae in Europe and other parts of the world (see references of Skuja 1925–1965). Israelson (1942) studied freshwater red algae in Sweden and compiled all older records in his work. Starmach worked with rhodophytes (especially in Poland) from 1952 to 1985 and also developed an identification key for the series Flora Słodkowodna Polski entitled 'Phaeophyta – Brunatnice / Rhodophyta – Krasnorosty' (Starmach, 1977). During the same period, Reis published numerous works on red algae of Portugal, describing several new *Batrachospermum* species (see reference list). Later, research activities on European freshwater red algae decreased although numerous articles were published (e.g. Alvarez Cobelas 1984, Cambra 1985, 1987, Aboal 1989, Sabater et al. 1989, Kwandrans & Eloranta 1994, Lokhorst 1994, den Hartog & Lokhorst 1995, Eloranta & Kwandrans 1996, 2002, 2007, Sheath & Sherwood 2002, Egidos & Aboal 2003, Kučera & Marvan 2004, Kostkevičienė & Laučiūtė 2005, Wołowski et al. 2007).
Apart from the identification key by Starmach (1977) including all taxa known worldwide, a new key 'Freshwater Red Algae of the World' with more recent results and records of freshwater red algae was published by Kumano (2002). Eloranta & Kwandrans (2007) published a brief identification guide to European taxa, including particularly those found in Finland.

Origin and diversity

Red algae are one of the oldest groups of eukaryotic algae. According to the latest results based on molecular aspects, they have a common ancestor with the green line including Streptophyta and Chlorophyta, but they are older (Eisen & Fraser 2003). Most red algae are marine, and estimates of the real species vary from 2540 (Dring 1982) to 6000 (Pritchard & Bradt 1984). According to Skuja (1938), freshwater red algae developed in two ways. The first group originates from freshwater cyanoprokaryotes, and the second group developed as secondary invaders from marine environments. This accounts for the preponderance of non-red Rhodophyta

in freshwaters, which have retained phycocyanin as dominant accessory pigment (secondary invaders have phycoerythrin as the major accessory pigment). Interestingly, most freshwater rhodophytes occur in rivers and streams rather than in lakes and ponds. Sheath (1984) stated, that the alternation of incoming irradiance through tree canopy might be a key factor affecting the adaptive radiation of primary freshwater species. Besides, the inability to use hydrogen carbonate is frequently found in freshwater taxa and therefore the specimens are restricted to sites with high carbon dioxide supply (Graham et al. 2009).
Skuja (1938) estimated that 5.7% (ca. 200) out of 3500 (according to Lee 1999, ca. 4000) red algal species can be found in freshwaters. Sheath (1984) listed 178 freshwater red algal taxa. The number of freshwater taxa has changed since 1984 due to descriptions of new taxa and combination/retraction of some older taxa. About one third of freshwater red algal taxa have been documented for Europe (64 species are considered in this book). The taxonomy is still in a state of flux, mainly due to new molecular data. Moreover, the interest in the study of freshwater red algae has increased in Europe during the last ten years; thus, several new records can be expected in the near future.

Classification

The classification of Rhodophyta is still not resolved. According to Christensen (1980), the phylum Rhodophyta comprises a single class Rhodophyceae. Starmach (1977), following Kylin (1956), divided the red algae into two classes: Bangiophyceae and Florideophyceae. Similarly, Sheath (1984) separated freshwater red algae into these two classes. According to Garbary & Gabrielson (1990), only one class of Rhodophyceae should be recognised because the inter- and infra-ordinal relationships between taxa belonging to the classes Bangiophyceae and Florideophyceae are currently unresolved. Kumano (2002) followed this argumentation, which is also accepted in the current key.
The classification system of red algae is under permanent development to reflect true phylogenetic relationships based on molecular studies and organelle ultrastructure. Recently, within a single class Rhodophyceae, two subclasses were proposed: Bangiophycidae, which appears to be polyphyletic, and Florideophycidae, which forms a monophyletic clade (see Kapraun 2005). Yoon et al. (2006) suggested a division of Rhodophyta into two subphyla, the Cyanidiophytina with a single class Cyanidiophyceae, and the Rhodophytina comprising the six classes Bangiophyceae, Compsopogonephyceae, Florideophyceae, Porphyridiophyceae, Rhodellophyceae and Stylonematophyceae.
The most characteristic features of red algae are: chlorophyll *a*, phycobiliproteins located in phycobilisomes (red phycoerythrin, blue phycocyanin and allophycocyanin), as well as chloroplasts lacking an external endoplasmatic reticulum and showing unstacked thylakoids. The storage product is floridean starch. Flagellas and centrioles are lacking even during reproduction. Rhodophyta generally show a diplohaplontic life cycle; sometimes a triphasic alternation of generations occurs, and pit-plugs are present in most orders (van den Hoek et al. 1995).

Cytological characteristics

The cell walls of red algae consist of cellulose microfibrils and an amorphous matrix containing sulphated galactan polymers and mucilages (mainly agar and

carrageenans, which are economically important in some marine taxa). This composition provides strength against physical stress, e.g. in fast running waters.
Conspicuous pit-plug connections between adjacent cells are a characteristic feature of many red algae and can be recognised already under a light microscope. At least seven different pit-plug connection types are known; they yield valuable taxonomic information. Some rhodophyte groups have plug cores, others feature structures filled with dense proteinaceous material. The function of pit-plugs is still uncertain, although they may act as structural reinforcements.
Chloroplasts vary in shape from discoid to stellate and to lobed forms. Sometimes, chloroplasts contain central pyrenoids; their function is still unknown. The storage product floridean starch (polysaccharide) is produced within the cytoplasm close to the chloroplast envelope, not in the chloroplasts as is the case in green algae. Chloroplasts have only two membranes (lacking an envelope of the endoplasmic reticulum), and the thylakoids are not stacked but separated (van den Hoek et al. 1995).
Red algae contain chlorophyll *a* (Jeffrey 1980). Earlier literature also mentioned chlorophyll *d* (Manning & Strain 1943, Smith & Benitez 1955, Strain 1958), but this never was confirmed. Thus, chlorophyll d may represent a degradation product (Gantt 1990). Some carotenoid pigments have been reported in red algae, including α- and β-carotene as well as cryptoxanthin, lutein, zeaxanthin and antheraxanthin (Strain 1958, 1966, Liaaen-Jensen 1978, Bjørnland et al. 1984, Schagerl & Donabaum 2003, Schubert et al. 2006). Apart from chlorophyll *a* and carotenoids, red algae also contain phycobiliproteids (red phycoerythrin, blue phycocyanin and allophycocyanin; e.g. Gantt et al. 1979, 1986). These commonly mask chlorophyll, giving the plants the characteristic bluish, reddish, violet or olive-green colour. Phycobilins are located in small bodies termed phycobilisomes on the surface of thylakoids (Gantt & Conti 1966, Gantt 1977). Interestingly, about 80% of freshwater red algae are not red (Sheath 1984), because they either lack phycoerythrin (e.g. *Batrachospermum turfosum*, Rider & Wagner 1972; *Cyanidium caldarium*, Diner 1979; *Porphyridium aerugineum*, Gantt et al. 1979) or, more frequently, phycocyanin masks phycoerythrin (Hirose & Kumano 1966, Gantt et al. 1979, Honsell et al. 1984).

Molecular studies

The recent development of molecular studies has fundamentally altered the traditional classification. Molecular data are used to gain insight into phylogenetic relationships between different levels of taxa and populations (e.g. Necchi Jr. & Entwisle 1990, Müller et al. 1997, 2001, 2002, Vis et al. 1999 Entwisle et al. 2009). More recently, research on molecular studies has been conducted mainly in North America, Australia and Japan (Vis & Sheath 1999, Vis et al. 1998, 2001, 2008 Vis & Entwisle 2000, Vis et al. 2001, Kumano 2002, Hanyuda et al. 2004, Sherwood et al. 2008, Entwisle et al. 2009). Some European rhodophyte populations are included in comparative studies (*Bangia*: Müller et al. 1998, *Balbiania*: Sheath & Müller 1999 and some species of *Batrachospermum*: Vis et al. 1998, Stewart & Vis 2007, *Hildenbrandia*: Sherwood & Sheath 2000, Sherwood et al. 2002).
DNA sequence analysis of the large subunit of the ribulose-bisphosphate carboxylase gene, (*rbc*L) and 18S rRNA gene data, together with morphometric results, showed a strong relationship between European and North American freshwater populations of *Bangia atropurpurea* (Müller et al. 1998). Contrary to the previous

opinion stated by Geesink (1973), *Bangia atropurpurea* should be kept separated from the marine species *B. fuscopurpurea*. Molecular data clearly supported the idea of the single ancestor for the European freshwater populations of the genus *Hildenbrandia*, and showed that there were no big differences between marine and freshwater *Hildenbrandia* species (Sherwood et al. 2002). Harper and Saunders (2002) studied the phylogeny of the order Acrochaetiales using combined small and large subunits of nuclear rDNA sequence data. This led to a re-classification of this order. Similarly, well-supported molecular findings were the reason for establishing the new orders Balbianiales (Sheath & Müller 1999) and Thoreales (Müller et al. 2002).

Sequencing the *rbc*L region and 18S rDNA of certain Batrachospermales showed the taxonomic status of genus *Sirodotia* species with respect to *Batrachospermum* and within the Batrachospermales (Vis et al. 1998, Vis & Sheath 1999). This is contrary to the previous concept that the genera *Tuomeya*, *Sirodotia* and *Nothocladus* should be treated as sections within the genus *Batrachospermum* (Necchi Jr. & Entwisle 1990). According to Vis & Sheath (1999), the *rbc*L spacer region may even provide evidence for a closer phylogenetic relationship between *Sirodotia* and *Batrachospermum gelatinosum*. The molecular data were recently used to establish a phylogenetic concept within the Batrachospermales (Vis & Sheath 1997, 1999, Müller et al. 1997, Vis et al. 1998, 1999, Kapraun et al. 2007, Stewart & Vis 2007, Sherwood et al. 2008). The final results were summarised by Entwisle et al. (2009).

Reproduction and life cycles

Freshwater red algae have several different reproduction types. All the reproductive cells in Rhodophyta are naked spherical protoplasts without flagellas and basal bodies. They develop in sporangia or gametangia.

Unicellular coccal forms grow asexually via simple cell division. Some multicellular forms (filamentous and pseudofilamentous forms) have an asexual reproduction mode by monospore formation. In *Chroodactylon* and *Bangia*, monospores are released from normal vegetative cells. In the genus *Compsopogon* some larger vegetative cells split by oblique cell division, forming a monosporangium. In the genera *Audouinella*, *Bangia* and in some *Batrachospermum* species, ovoid and enlarged monospores are produced in monosporangia formed by apical cells of vegetative branches.

Sexual reproduction is oogamous. Haploid gametophytes produce male nonmotile, small spermatia in spermatangia (one spermatium in each spermatangium) and female reproductive cells termed carpogonia. A carpogonium is a typically elongated cell with an enlarged base and an extended, colourless process known as a trichogyne. For sexual reproduction, a spermatium attaches to the trichogyne. Afterwards, its nucleus migrates via the trichogyne to the carpogonium nucleus located at the basal area, where the fusion takes place.

The life cycle of the freshwater genus *Audouinella* is triphasic, showing isomorphic gametophytes and tetrasporophytes. After fertilisation of carpospores, a small diploid carposporophyte develops, which remains attached to the haploid gametophyte. Carpospores produced by ripe carposporophytes are released into the water, settle and germinate. This diploid generation termed tetrasporophyte forms tetrasporangia, in which meiosis takes place. This yields haploid tetraspores, which are released, geminate and eventually develop to gametophytes (Drew 1936, Necchi Jr. et al. 1993, Sheath 2003).

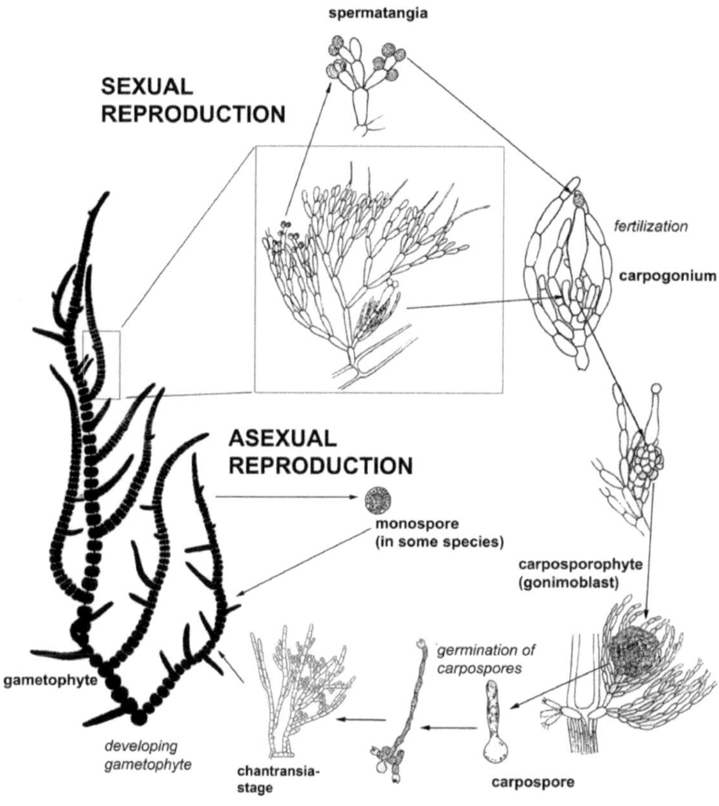

Fig. 1. The life cycle of *Batrachospermum* (modified from Walter 1961).

In the order Batrachospermales, the diplohaplontic life cycle consists of a macroscopic gametophyte phase and a microscopic filamentous *'Chantransia'* stage (Sheath 1984). The life cycle is similar to *Audouinella*, except for the tetrasporophyte, which is known as the *'chantransia'* stage. Here, meiosis takes place in apical cells (no tetraspores are produced), which develop directly to macroscopic gametophytes (Sheath 1984) (fig. **1**).
The *Lemanea*-type of life history is similar to the life history of *Batrachospermum*, but the carposporophytes develop inside the tubular thallus (fig. **2**).
Sexual reproduction and life cycle phases are still not completely demonstrated for freshwater species in the orders Bangiales, Compsopogonales and Hildenbrandiales.

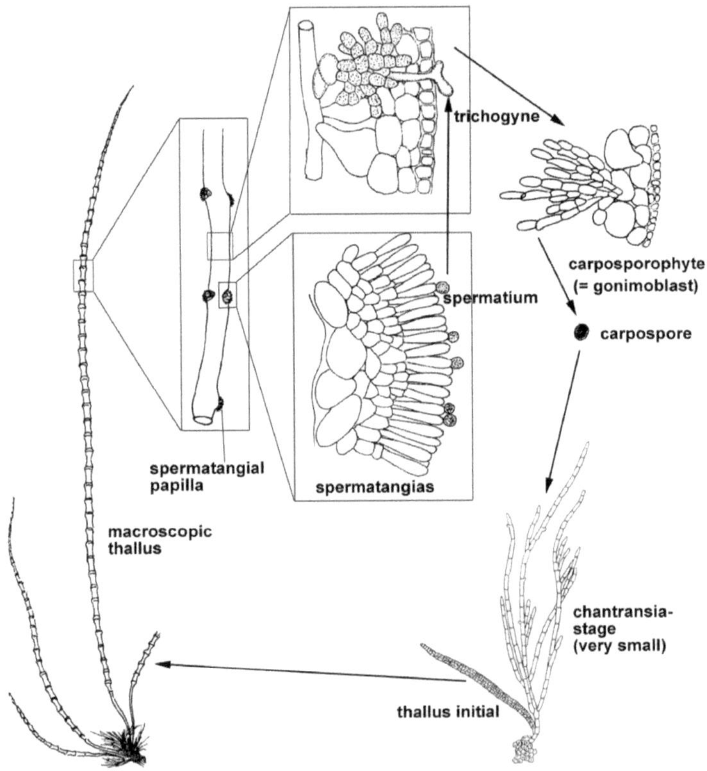

Fig. 2. The life cycle of *Lemanea* (modified from Sirodot 1872 in Starmach 1977).

Growth forms

Freshwater red algae have diverse growth forms. The simplest are **unicells** (6.3 % of freshwater taxa; Sheath 1984) living either planktonically or attached on moist substrata forming mucilaginous, loose colonies (e.g. *Rhodospora*, *Chroothece* and *Porphyridium*). Some genera form **pseudofilaments** *(Chroodactylon*, *Kyliniella* and *Phragmonema)*. One North American genus *(Boldia)* occurs as **monostromatic sacs**, much like the green alga *Enteromorpha*. Most freshwater rhodophytes, however, are unbranched *(Bangia)* or branched **filaments** (74.7%; Sheath 1984). Some taxa are uniseriate (e.g. *Audouinella*) without cortication, while others such as the common *Batrachospermum* and *Sirodotia* have the main axis composed of axial cells with cortical filaments enveloping the main axis. In the genera *Thorea* and *Nemalionopsis*, **filaments** are **compacted** together and form cord-like thalli. Both *Lemanea* and *Tuomeya* form **pseudoparenchymatic,** cartilaginous thalli (18.4% of freshwater taxa; Sheath 1984) with a distinct cortex layer. The genus *Hildenbrandia* can easily be recognised by its red **crustose thalli** growing on hard substrata. The crusts are constructed by short, erect, closely packed filaments.

Structure of thallus in the families **Batrachospermaceae** and **Lemaneaceae**

The main thallus is usually rich in branching (fig. **3**), which could be irregular, dichotomic or pseudodichotomic. The main axis is composed of axial cells covered with cortical filaments running on the surface of the main axis. With the exception

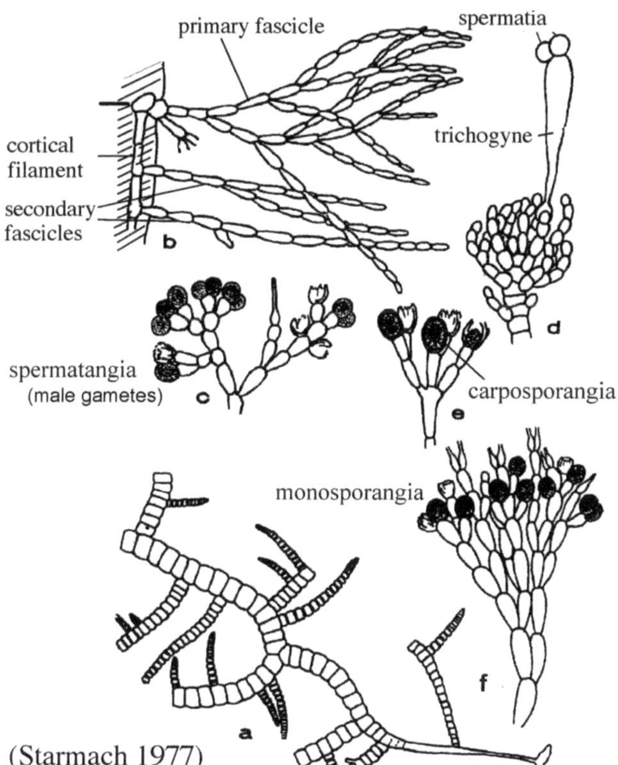

Fig. 3. *Batrachospermum*: **a** – Basal part of the thallus with compressed whorls; **b** – primary and secondary fascicles; **c** – spermatangia; **d** – carpogonium-bearing branch with bracts, carpogonium and its trichogyne, **e** – gonimoblast filaments with carposporangia, **f** – monosporangia terminal on fascicles.

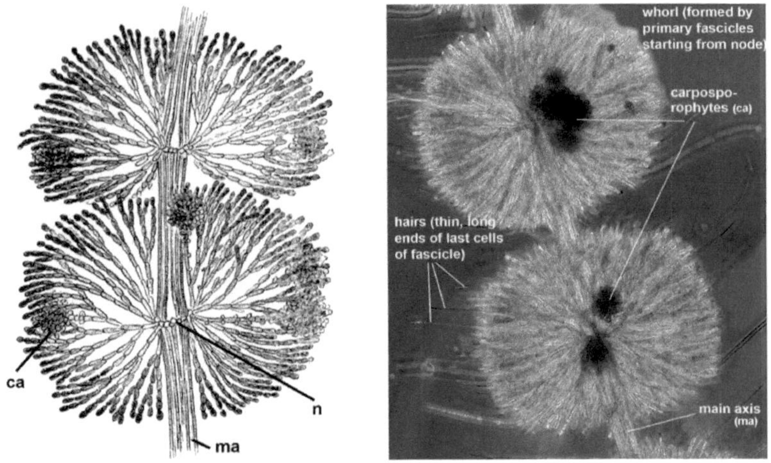

Fig. 4. *Batrachospermum*: Spherical whorls with carposporophytes (ca), main axis (ma) with cortical filaments and nodes (n).

of the genus *Tuomeya* (here cortical filaments always make a dense cortex around the axial cells), the main axis can be without any cortication in some parts of the thallus.

So-called primary fascicles start radially from the axial nodes forming spherical or barrel shaped whorls (fig. **4**). In many taxa, secondary fascicles arise from cortical filaments, which can be shorter or the same length as primary fascicles; they are usually unbranched. The outermost cells of fascicles may have terminal hairs, which vary in abundance and length. The shape and size of cells may vary along the fascicle, as also the number of the cell storeys might be different.

The carpogonium-bearing branches arise from periaxial cells and/or from fascicle cells or cortical filaments. In some species, carpogonium-bearing branches are can not be distinguished from other fascicles whereas in others, those branches are differentiated (carpogonial branches) and have a definite number of unmodified cells bearing a carpogonium (fig. **5**). The carpogonium can be surrounded by shorter or longer filaments (bracts). The carposporophyte (gonimoblast) develops from the fertilized carpogonium, and is formed by short filaments (gonimoblast filaments) with carposporangia on top (fig. **5**). In the genus *Sirodotia*, the carposporophyte is a simple, creeping filament along the main axis on which carposporangia develop. Finally, carpospores are released, germinate and develop to new chantransia-stages of the plant.

In the family Lemaneaceae thalli are stiff pseudoparenchymatous threads. In the centre of the thread, long axial cells are located. Periaxial basal cells of fascicles grow radially (lateral filament or ray-cells) and outer cell layers smaller in size form a compact outer cortex (fig. **6**).

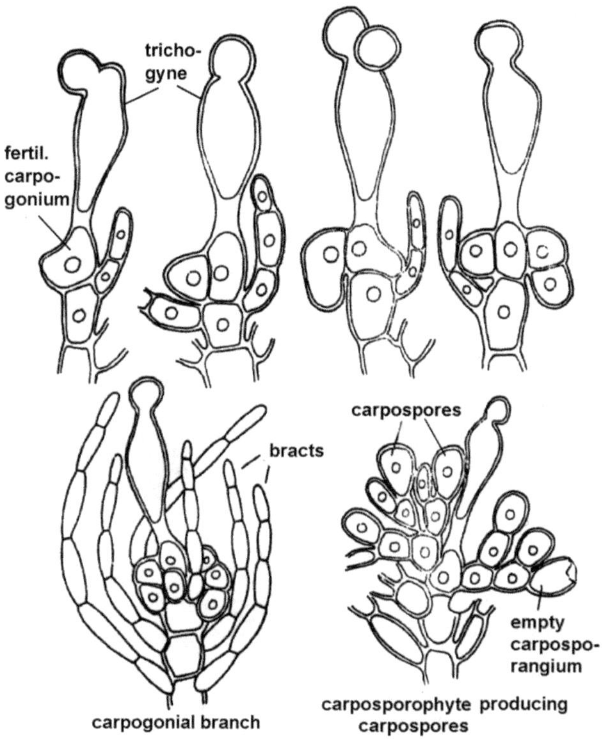

Fig. 5. *Batrachospermum*: Steps in the development of the carposporophyte after fertilization of carpogonium (modified from Kylin 1930 in Starmach 1977).

Ecology and habitats of freshwater taxa

Generally, more taxa are found in the tropics than in the temperate zone. *Compsopogon* and *Thorea* are typical warm-water genera with their main distribution area in the tropics (Kann 1978, Kumano 1980). However, they are also known from temperate European environments during summer months (Descy & Empain 1974, Tomas et al. 1980, John et al. 2002, Žakova et al. 2010). Sometimes, *Compsopogon* can also be found in aquaria. According to studies conducted in North America, maximum growth and highest biomass of freshwater red algae occur from late autumn to early summer (Sheath 1984, Sheath & Burkholder 1985). Similar results have been recorded from the British Isles (Holmes & Whitton 1981). These findings, however, cannot be generalised. According to Starmach (1977), *Batrachospermum* may exist throughout the year if the water temperature remains below

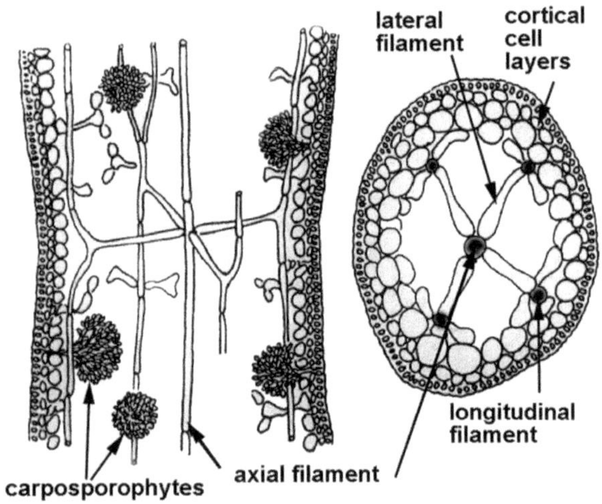

Fig. 6. Structure of *Lemanea* thallus (modified from Starmach 1977).

15 °C. In the Northern Temperate Zone of Europe, insolation is very low during the winter months; therefore, many smaller creeks and rivers are covered with ice and snow and sometimes frozen down to the bottom, resulting in a main occurrence of rhodophytes from March to October (Kylin 1912, Israelson 1942, Starmach 1984a, Eloranta & Kwandrans 1996).

Small coccal forms commonly grow on moist places such as soil, rocks or in peat bogs (e.g. taxa from the genera *Cyanidium*, *Rhodospora*, *Chroothece* and *Porphyridium*). *Porphyridium* favours sites with high nutrient concentrations. *Galdieria* and *Cyanidium* are found in acid hot springs. Small filamentous forms of the genera *Chroodactylon* and *Kyliniella* are litho- and epiphytic and occur in standing waters. *Balbiania* grows epiphytically on *Batrachospermum gelatinosum* and *B. keratophytum* (Starmach 1977).

Running waters are the typical habitats for many freshwater red algae, but some genera such as *Batrachospermum* and *Hildenbrandia* may also be found in both running and standing waters (e.g., *B. gelatinosum*, *B. keratophytum* and *B. turfosum*).

Ecological studies on European freshwater rhodophytes are rather scarce. According to extensive investigations from North America, species cover broad ecological ranges, for example with regard to current velocity (Sheath 1984). Similar results were also obtained from Finnish rivers (Eloranta & Kwandrans 1996). Particularly in small rivers, current velocity may vary considerably, e.g. after rains and between microhabitats. Apart from growing on rocks, some genera also occur epiphytically in running waters. This includes *Audouinella*, which can be found on water mosses but is mainly attached to *Lemanea*. *Sirodotia* grows on submerged wood, stones and on *Batrachospermum*.

Some taxa prefer hard waters with pH > 7.0 originating from calcareous catchment areas, among them *Bangia*, *Chroodactylon*, *Hildenbrandia* and *Thorea*, whereas many taxa prefer slightly acidic, soft waters (Sheath 1984, Eloranta & Kwandrans 1996). In the list of the alkaline water group (pH 7.0–8.0), Israelson (1942) lists *Hildenbrandia* as well as *Batrachospermum anatinum*, *B. atrum* and *B. helminthosum*. The highest frequency of occurrence in soft waters (specific conductivity $<100\ \mu S\ cm^{-1}$) in America was found for *Audouinella*, *Batrachospermum*, *Lemanea* and *Tuomeya* (Sheath & Hambrook 1990). The first three genera are rather common in the Nordic countries, where waters are generally soft and often at least slightly acidic, and in many cases also brownish due to dissolved humic substances. *Batrachospermum turfosum*, which mainly inhabits lakes and peat bogs, can survive from acidic to alkaline conditions, indicating a wide tolerance of many species for pH range (Israelson 1942, Eloranta & Kwandrans 1996). According to Sheath & Hambrook (1990), the frequency of occurrence of freshwater rhodophytes is negatively related to water pH, which can be explained by the carbon supply. At least some taxa are able to use only free carbon dioxide as an inorganic carbon source (Raven & Beardall 1981). As bicarbonate and carbonate dominate in alkaline waters, some species are either restricted to acidic water bodies, or at least some water flow is needed to replenish free CO_2 resources (Raven et al. 1982).

Light requirements of freshwater rhodophytes vary considerably between taxa and are correlated with intensity and mainly light quality (Hellebust 1970, Whitton 1975, Sheath 1984, Sheath & Burkholder 1985). Some taxa, such as most of the *Lemanea* species, occur in open rivers that experience high insolation (Kremer 1983), whereas euryphotic species occur at both high and low light conditions (Rider & Wagner 1972). Nonetheless, most freshwater red algae seem to favour shadowed places (Sheath & Hambrook 1990, Necchi Jr. 2005), e.g. river sectors with dense canopy cover or brownish colour. Tree canopy may reduce solar radiation by 90% (Sheath & Hambrook 1990). Apart from reducing solar radiation, tree canopy changes the light spectrum reaching the water, thereby mainly reducing red wavelengths (Federer & Tanner 1966 in Sheath & Hambrook 1990). Also in dystrophic waters typical of northern Europe, red wavelengths dominate in the water. This may compensate for the effect of tree canopy and explain, at least partly, the rather high abundance of freshwater red algae in those countries.

Rhodophytes versus trophy and saprobity

Few recent studies are available on tolerances of rhodophytes to different ecological factors. In general, freshwater rhodophytes occur in a wide range of nutrient levels (Sheath & Hambrook 1990), but most prefer rather oligotrophic waters with low turbidity and an environment offering clean substrata for attachment. In eutrophic rivers, substrata (mainly rocks) are often covered with silt and dense biofilms. Rott et al. (1999) developed a classification for the trophic preferences of the running water red algae. In this classification, only few taxa are classified as eutrophic (eu) or meso-eutrophic organisms (me-eu). These include: *Audouinella hermannii* (me-eu), *Bangia atropurpurea* (me-eu), *Batrachospermum gelatinosum* (eu), *Lemanea fucina* (eu) and *Thorea hispida* (eu). *Audouinella pygmaea* is mesotrophic. All the remaining taxa are classified as oligotrophic or ultra-oligotrophic. In the Nordic countries, *Audouinella* spp. covers a wide ecological spectrum, including oligotrophic and eutrophic waters (Eloranta & Kwandrans 1996). Similarly, in Finnish waters, *Batrachospermum gelatinosum* and *Lemanea fucina* are the most common taxa in their genera, mostly inhabiting oligotrophic waters,

whereas *B. helminthosum* (syn. *B. virgatum* Sirodot nom. illeg.) and *Lemanea mamillosa* occur also in eutrophic Finnish rivers and creeks.
Saprobity systems are developed to describe mainly organic pollution of waters. Fjerdingstad (1964, 1965), Backhaus (1973) and Sladeček (1973) regarded some freshwater red algae as indicators in their saprobity systems. Most of the defined rhodophyte taxa were classified to be indicators of unpolluted katharobic or oligosaprobic waters. Some taxa, for example *Audouinella chalybea*, with wider ecological tolerance, occurred from katharobic to even α-mesosaprobic levels.
Rott et al. (1997) provided the saprobity classification for 24 red algal taxa. In their list, 4 taxa were classified as saprophobic (saprobic value $S < 1.3$): *Balbiania investiens*, *Batrachospermum helminthosum*, *Chroothece richteriana* and *Paralemanea catenata*. *Audouinella hermannii* and *A. chalybea* were classified as saprophilous ($S = 2.2$–2.5), the remaining were slightly or moderately tolerant ($S = 1.4$–2.1) against organic load.

Sampling and preservation

Most of the European freshwater red algal taxa have their maximum abundance in cold waters, making collection easiest during the cold season or in generally cold waters. There are, however, a few exceptions for taxa that live in hot springs or are typical of subtropical or tropical environments.
Most of the freshwater red algal taxa occur in streams and rivers, although some can also be found in ponds and lakes. Many taxa prefer smaller, first to fourth order streams with tree canopy or/and dystrophic water. Some taxa, for example *Lemanea*, are attached to substrata in large rivers experiencing high insolation. Other taxa of the family Batrachospermaceae typically grow near the shoreline of river channels, where current speed is decreased.
For sampling in running waters, 20–50 m of a flowing stretch should be examined using a view box with glass or Plexiglas bottom (fig. 7). Samples can be grabbed using long forceps (e.g. *Lemaneas* or other bigger forms) or using a large (50–100 ml) plastic pipette. *Lemanea* should be carefully removed as to include its basal parts. The best way to collect crustaceous forms is to take them together with the substratum. In the laboratory, material can be scraped off with a sharp knife or a razor blade.
Physical and chemical measurements, for example with portable meters, are recommended. Basic ecological data include water pH, temperature, conductivity, turbidity and current velocity.
Samples should be stored in wide bottles and preserved in 3% formaldehyde or in 2.5% buffered glutaraldehyde (Sheath 2003). All visible features such as colour, size, and thallus structure should be recorded in the field. To store the collection over longer periods, 70% ethanol together with 5% glycerine (to avoid accidental drying out) and some thymol crystals can also be used (Starmach 1977). Herbarising specimens can be done for crusts or larger forms *(Batrachospermum, Lemanea, Thorea)* but may result in morphological changes, especially in small organs needed for identification.
For molecular investigations, separate specimens should be taken, cleaned and put into a small plastic bags or tubes, and desiccated with pre-dried silica gel granules. Those granules should be changed after one day. Samples should be stored in a deep-freezer until analysis.

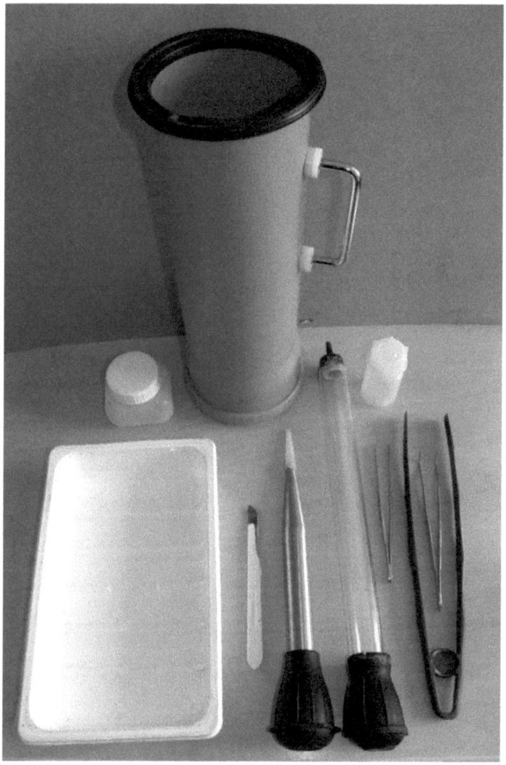

Fig. 7. Sampling tools.

According to Starmach (1977) freshwater rhodophytes could be kept in cultures but, e.g., the whole life cycle growth of *Batrachospermum* is not completed in the artificial environments.

Glossary

Antheridium (pl. antheridia) (= spermatangium) – organ forming male gametes
Antheridial papillae – nodes in *Lemanea* containing spermatangia (= spermatangial papillae)
Apex (pl. apices) – tip of the filament
Apical growth – growth by division of one or more apical cells of a multicellular thallus
Assimilatory filament – special filament type in the Thoreaceae consisting of pigmented cells
Auxiliary cell – cell that receives the zygote nucleus after fertilisation; the nutritive auxiliary cell provides nutrients for the developing carposporophyte; the generative auxiliary cell enables development of gonimoblast filaments
Axial cell – cells forming the main filament
Bract filament – filament or fascicle formed by cells of the carpogonial branch (= involucral filaments)
Branchlet – small secondary branch (see fascicle)
Carpogonium (pl. carpogonia) – female gametangium
Carpogonial branch – short branch directly below the carpogonium
Carpogonium-bearing branch – side branch which may contain several carpogonia and also several carpogonial branches
Carposporangium – carpospore-producing sporangium
Carpospore – diploid spore produced by the carposporangium
Carposporophyte (gonimoblast) – diploid generation derived from the zygote, and which forms carpospores via the gonimoblast filaments; always attached to the female gametophyte
Chantransia stage – diploid generation in the Batrachospermales developing from germinating carpospores. Homologous to the tetrasporophyte, but no tetraspores are produced during meiosis.
Clavate – wedge-shaped
Contorted – twisted
Cortex – layer of cells enveloping axial filaments or cells
Cortical cells – cells forming the cortex
Cortical filament – filament running from the node along the axial cells
Cruciate – cross-shaped
Dichotomy – branching of a thallus into two equal branches
Dioecious – (= heterothallic) male and female gametangia on separate gametophytes
Endogenic – derived internally
Endospore – spore formed within cell
Fascicle – branchlets arising from the main axis
Filiform – thread-like
Foliose – leaf-like
Fusiform – spindle-like
Gametangium – organ forming gametes
Gametophyte – haploid generation that forms the gametes
Gemma – small asexual reproductive structure, that becomes detached from the parent thallus and develops into a new individual
Generative auxiliary cell – cell forming the gonimoblast filaments

Gonimoblast – see carposporophyte
Gonimoblast filament – short filaments bearing carposporangia
Heterothallic – (= dioecious) male and female gametangia on separate gametophytes
Homothallic – (= monoecious) male and female gametangia on the same gametophyte
Hypogenous cell – cell situated below the carpogonium
Indeterminate – without exact, special structure
Intercalary – inserted between, e.g., cells, filaments or tissues
Internode – part of the axis in between nodes
Involucral filament – filament formed by cells of the carpogonial branch (= bract filament)
Involute – curled, rolled inwards
Medulla – inner part of algal thallus, usually composed of packed colourless filaments
Microgonidia – small asexual spores in the family Phragmonemataceae, order Goniotrichales
Microspore – small asexual spores in some *Compsopogon* taxa
Monoecious – (= homothallic) having male and female gametangia on the same gametophyte
Monosporangium – a sporangium that forms a monospore
Monospore – asexual spore for dispersal
Multiaxial – many similar axial filaments arranged in parallel
Multiseriate – more than one row of cells
Node – part of thallus that bear branches
Nodular cell – cell of a fascicle arising from a node
Obconical – cone-shaped with the broader end ahead
Parenchyma – thin-walled living cells that produce true tissues by division
Parietal – peripheral
Pedicellate (= pedunculate) – with a stalk-like base
Pericentral cells – (**periaxial cells**) basal cells of the primary fascicles
Phycobiliprotein – water-soluble blue-green or pink pigment
Pit connection – area between two red algal adjacent cells consisting of an aperture, a plug, and a plug cap
Pit-plug – structure of pit connection consisting of core, cap membrane enveloping core, and cap layer
Polyoecious – male and female gametangia on the same **and** on separate gametophytes
Primary fascicle – fascicle initiating from the node
Procarp – association of carpogonium after fertilisation and auxiliary cells
Prostrate – horizontally creeping
Pseudocarposporophyte – type of carposporophyte in *Batrachospermum pseudocarpum* consisting of indeterminate gonimoblast filaments
Pseudodichotomous – branched with the strong development of a lateral branch to resemble the original axis
Pseudofilament – thread of cells incidentally arranged in a linear series, not a true filament
Pseudoparenchyma – tissue-like structure formed by complex of tightly interwoven filaments
Pyramidal – shape of pyramid

Pyriform – pear-shaped
18S rRNA gene – sequences of nuclear ribosomal DNA gene small subunit
rbcL gene – sequences of the chloroplast ribulose-1,5-biphosphate carboxylase/oxygenase gene large subunit
Ray cell(s) – a type of periaxial cells found in *Lemanea* and *Paralemanea*
Secondary fascicle – fascicle initiating from the cortical filament
Sessile carposporophyte – carposporophyte attached directly to the main axis
Spermatangial papillae (antheridial papillae) – papillae at nodes of *Lemanea* producing spermatia
Spermatangium – male gametangium forming one non-motile spermatium
Spermatium – male, non-motile gamete, no flagellas present
Sporangium – spore-producing organ
Spore – asexual reproductive cell
Sporophyte – diploid plant that forms spores
Tetrasporangium – a sporangium producing four tetraspores, usually through meiosis
Tetraspore – spore formed in tetrasporangium through meiosis
Tetrasporophyte – diploid phase forming tetraspores
Trichogyne – long colourless tip of a carpogonium that receives the spermatium
Truncate – flat at the top, flatly rounded
Uniaxial – single axial filament
Uniseriate – having a single row of cells
Whorl – the bunch of primary fascicles running from the same node
Zygote – product of the fusion of two gametes

Systematic part

Phylum **Rhodophyta**

Class **Rhodophyceae** Ruprecht

Key to orders:

1a Cells unicellular, free-living or embedded in mucilaginous matrix forming irregular colonies . . . . **1. Porphyridiales**
1b Cells forming uni- or multiseriate filaments or threads, or pseudoparenchymatic or crustaceous thalli **2**
2a Thallus composed of diffuse uniseriate filaments, which in some taxa arise from a pseudoparenchymatous base . . . . **2. Goniotrichales**
2b Thallus composed of either unbranched or branched uni- or multiseriate filaments or showing pseudoparenchymatic or crustaceous structures **3**
3a Thallus uniseriate, branching filaments corticated with small cells covering central axial cells; reproduction by monospores or by microspores . . . . **3. Compsopogonales**
3b Thallus different **4**
4a Thallus branched, uni- or multiseriate with apical growth **5**
4b Filaments unbranched, uni- or multiseriate with intercalar growth, sexual reproduction not known, asexual reproduction by monospores . . **4. Bangiales**
5a Thallus monosiphonous, branched filaments without mucilaginous envelope **6**
5b Thallus uniaxial with cortication, multiaxial or crustaceous **7**
6a Erect filaments forming tufts attached to substratum . . . . **5. Acrochaetiales**
6b Small creeping filaments growing on *Batrachospermum*, spermatangia born on top of long differentiated cells . . . . **6. Balbianiales**
7a Thallus composed of a main axis with an axial cell row and whorls of short branchlets or pseudoparenchymatic thalli are consisting of tubular tufts . . . . **7. Batrachospermales**
7b Thallus multiaxial or crustaceous **8**
8a Thallus multiaxial, lateral assimilatory filaments along entire axis . . . . **8. Thoreales**
8b Thallus crustaceous, forming red patches on stones . . . **9. Hildenbrandiales**

1. Order **Porphyridiales** Kylin 1937

Unicellular algae living in fresh waters, brackish waters and on various moist substrata. Cells are single or in colonies (e.g. *Porphyridium*). Asexual reproduction by binary division or by autospores; sexual reproduction unknown.

Important references: Starmach (1977); Kumano (2002); Sheath & Sherwood (2002)

Key to families:

1a Cells have one stellate chloroplast comprising a central pyrenoid; reproduction by binary division . . . . **1. Porphyridiaceae** (p. 18)
1b Cells contain one chloroplast without pyrenoid; reproduction by autospores . . . . **2. Cyanidiaceae** (p. 20)

1. Family **Porphyridiaceae** Kylin ex Skuja 1939

Unicellular, cells have one stellate chloroplast containing a single central pyrenoid; sexual reproduction unknown; asexual reproduction by binary division.

Key to genera:

1a Cell membranes mucilaginous, delicate, homogenous, cells spherical .... **1. Porphyridium**

1b Cell membranes layered, cells elliptical to elongated elliptical **2. Chroothece**

1. **Porphyridium** Nägeli 1849

Cell membranes delicate, cells embedded in a mucilaginous and homogenous matrix (not layered). All species favour eutrophic environments.

Key to the species:

1a Cells >8.5 µm in diameter **2**

1b Cells <8.5 µm in diameter **3**

2a Cells blood-red or brownish-red, 7–12(–15) µm in diameter, on wet soils or walls .... **1. P. purpureum**

2b Cells grey, 8.5–16.5 µm in diameter, scattered randomly in mucilage .... **2. P. griseum**

3a Cells blue-green, 4.8–8.5 µm in diameter single-celled or in small colonies among other algae in ponds and rivers .... **3. P. aerugineum**

3b Cells grey-green, olive-green or yellowish, 5.5–8.5 µm in diameter; occurs on thalli of *Phormidium* in runlets from dungheaps .... **4. P. sordidum**

Note: According to Sheath & Sherwood (2002), species are distinguished by chloroplast colour.

1. **Porphyridium purpureum** (Bory) Drew et Ross 1965 (plate **1**: fig. **1**)

Phytoconis purpurea Bory 1797, *Porphyridium cruentum* (Gray) Nägeli 1849

Cells spherical, blood-red to brownish red, 7–12(–15) µm in diameter with a single stellate chloroplast containing a central pyrenoid. Colonies irregular, often found as large mucilaginous patches on moist, shaded substrata.
Occurrence: Grows e.g., on moist walls with calcareous components and on wet soil. According to Starmach (1977), the main occurrence is from September to November. Common all over Europe.

2. **Porphyridium griseum** Geitler 1970 (plate **1**: fig. **2**)

Cells grey, 8.5–16.5 µm in diameter, randomly scattered in mucilage; chromatophore three-dimensional stellate with pyrenoid and starch grains.
Occurrence: Found in Lake Neusiedler See, Austria.

3. **Porphyridium aerugineum** Geitler 1923 (plate **1**: fig. **3**)

Cells blue-green, 4.8–8.5 µm in diameter; chromatophore stellate with a central pyrenoid. Either single cells or small colonies.
Occurrence: Occurs on moist soils and according to Starmach (1977) also in ponds and rivers. Europe: Great Britain.

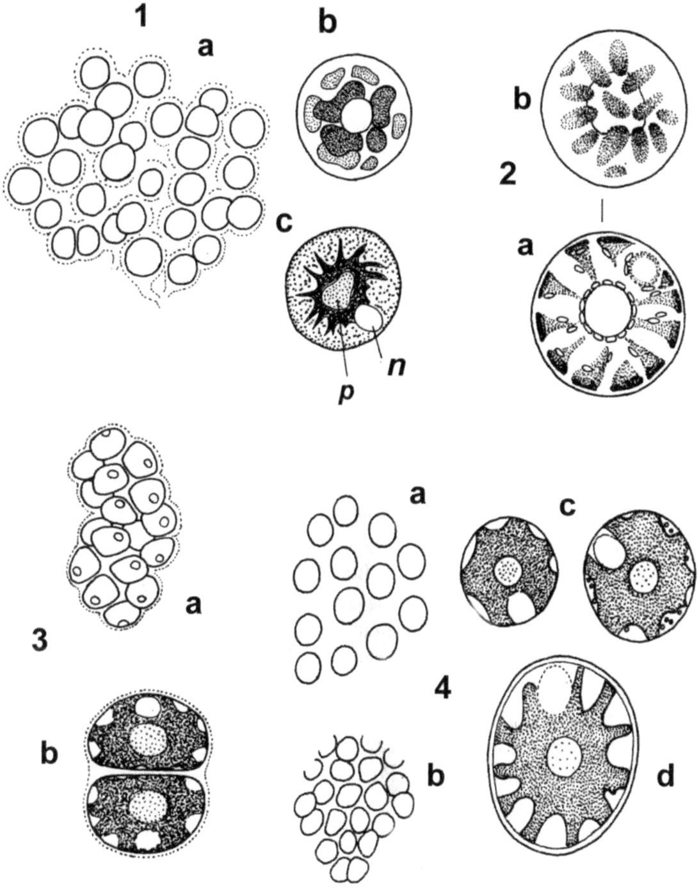

Plate 1: **Fig. 1.** *Porphyridium purpureum* (original author, Geitler 1932) **a** – part of colony; **b** – single cell; **c** – cross-section of cell: p – pyrenoid, n – nucleus. **Fig. 2.** *Porphyridium griseum* (original author, Geitler 1932): **a** – cross-section of cell; **b** – cell surface view. **Fig. 3.** *Porphyridium aerugineum* (original author, Geitler 1932): **a** – part of colony; **b** – dividing cell. **Fig. 4.** *Porphyridium sordidum* (original author, Geitler 1932): **a** – part of old colony; **b** – part of young colony; **c** – young cells; **d** – resting cell.

4. Porphyridium sordidum Geitler 1932 (plate **1**: fig. **4**)

Cells grey-green, olive-green or yellowish, 5.5–8.5 µm in diameter; cell membrane very thin; colonies loose or clustered, usually in one layer.
Occurrence: Occurs on thalli of *Phormidium* in runlets from dungheaps. Europe: Found in Pinggau, eastern Austria.

2. Chroothece Hansgirg 1884

Petrovanella Kylin 1956

Colonies mucilaginous, irregular; cell membranes layered. Dimensions given without envelope.

Key to the species:

1a	Cells <10 µm in diameter	**2**
1b	Cells >20 µm in diameter, up to 44 µm long	**1. C. mobilis**
2a	Cells 6–10 µm in diameter, 15–18 µm long	**2. C. richteriana**
2b	Cells 5–7 µm in diameter, 9–15 µm long	**3. C. rupestris**

1. Chroothece mobilis Pascher et Petrova 1931 (plate **2**: fig. **4**)

Petrovanella mobilis (Pascher et Petrova) Kylin 1956

Cells without envelope 24 µm in diameter, up to 44 µm long, located on stalks coalescent with their bases to a mucilaginous mass. Cells have apertures at their tips by which mucus is excreted.
Occurrence: In peat-bogs.

2. Chroothece richteriana Hansgirg 1884 (plate **2**: fig. **2**)

Cells without envelope 6–10 µm in diameter, 15–18 µm long; thallus slimy, 2–4 mm thick, blue-green, becoming brownish, yellowish to orange by age. Cell envelope distinctly layered, in older cells unilaterally developed.
Occurrence: On humid clay-rich soil at ± saline sites; Europe: Great Britain, Spain.

3. Chroothece rupestris Hansgirg 1884 (plate **2**: fig. **3**)

Cells without envelope 5–7 µm in diameter, 9–15 µm long; thallus shapeless, at first light-blue then orange-yellow. Chromatophore coarsely star-shaped.
Occurrence: On wet rocks; Europe: Czech Republic, Spain.

2. Family **Cyanidiaceae** Geitler 1933 emend. Chapman 1974

Unicellular, cells with one to more chloroplasts without pyrenoid; asexual reproduction by binary division or autospores.

Note: For discussion of genus separation see Albertano et al. (2000) and Kumano (2002).

Key to genera:

1a	Reproduction by binary cell division	**1. Cyanidioschyzon**
1b	Reproduction by autospores or endogenic cell division	**2**
2a	4 autospores formed	**2. Cyanidium**
2b	4–32 autospores formed	**3**
3a	Several parietal chloroplasts per cell; reproduction by endogenic cell division or by 4–32 autospores	**3. Rhodospora**
3b	Cells contain one, parietal chloroplast without pyrenoid; asexual reproduction by autospores	**4. Galdieria**

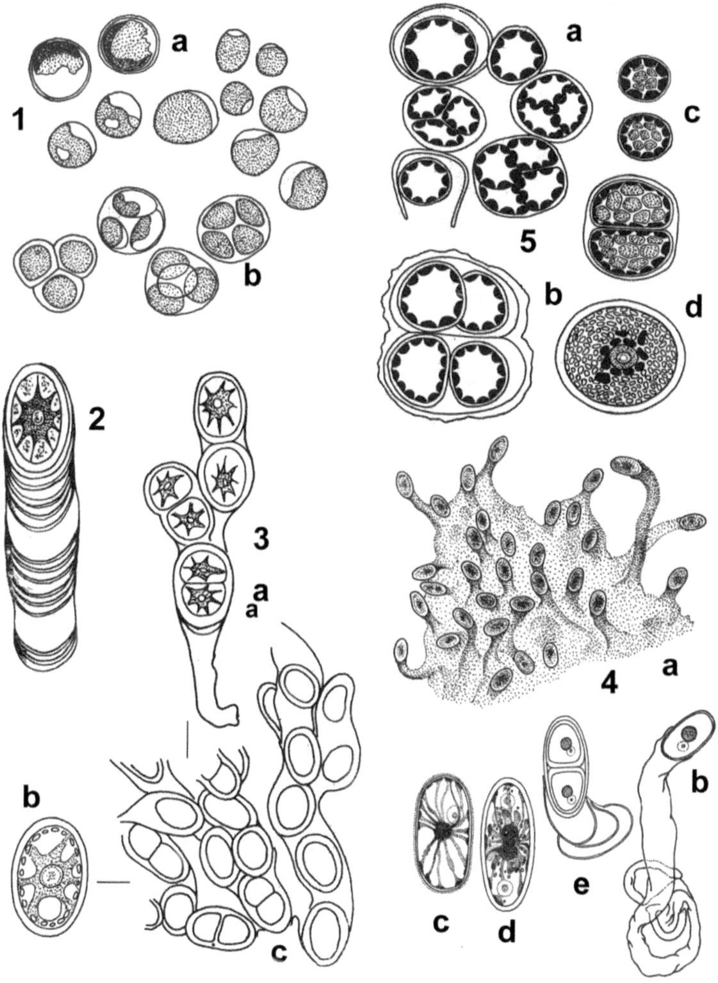

Plate 2: **Fig. 1.** *Cyanidium caldarium* (original author, Geitler 1932): **a** – vegetative cells; **b** – autospore development. **Fig. 2.** *Chroothece richteriana* (original author, Hansgirg 1884). **Fig. 3.** *Chroothece rupestris* (original author, Israelson 1949): **a** – young colony; **b** – cells with stellate chloroplast; **c** – growing colony. **Fig. 4.** *Chroothece mobilis* (original authors, Pascher & Petrov 1931): **a** – part of colony with cells at tops of mucilage stalks; **b** – single cell at the top of mucilage cord; **c** – single vegetative cell; **d** – single cell under division. **Fig. 5.** *Rhodospora sordida* (original author, Geitler 1932): **a** – part of colony showing autospore formation; **b** – group of cells; **c** – vegetative cells; **d** – resting cell.

1. **Cyanidioschyzon** De Luca et al. 1978 in NCU-3 1993

Cells with one ovoidal chloroplast without pyrenoid. Asexual reproduction by binary cell division. One species described.

Cyanidioschyzon merolae De Luca, Taddei et Varano 1978 (plate **4**: fig. **1**)

Cells blue-green, club-shaped, 3–4 µm long and 1–2 µm in diameter; one elongated chloroplast without pyrenoid. Cell walls very thin or invisible even at higher magnification. Before division is completed, chloroplast divides longitudinally thus causing a V-shape of the cell.
Occurrence: Occurs worldwide as thin layers on stones in acid hot springs; Europe: Italy.

2. **Cyanidium** Geitler 1933 in NCU-3 1993

Rhodococcus Hirose 1958, non *Rhodococcus* Hansgirg 1885

Cells with one ovoidal chloroplast, pyrenoids absent. Asexual reproduction by autospores. Monospecific genus.

Cyanidium caldarium (Tilden 1898) Geitler 1933 (plate **2**: fig. **1**)

Rhodococcus caldarius (Tilden 1898) Hirose 1958

Cells blue-green, spherical or rounded pyramidal, 4–6 µm in diameter. Autosporangium with 4 trihedrally arranged autospores (2–3 µm in diameter each).
Occurrence: Ecology and occurrence as for *Cyanidioschyzon*; worldwide distributed and living on stones in acid hot springs. Europe: Italy.

3. **Rhodospora** Geitler 1927

Cells with several parietal chloroplasts; occurring as irregular colonies on wet stones. Reproduction by endogenic cell division or by 4–32 autospores. Monospecific genus.

Rhodospora sordida Geitler 1927 (plate **2**: fig. **5**)

Cells up to 18 µm in diameter, chloroplasts brownish, olive green, green or violet brown, varying between cells.
Occurrence: On wet rocks; Europe: Austrian Alps, Spain, Sweden.

4. **Galdieria** Merola in Merola et al. 1982 in NCU-3 1993

Cells with parietal, lobed chloroplast without pyrenoid. Asexual reproduction with 4–32 autospores. Monospecific genus.

Galdieria sulphuraria (Galdieri 1899) Merola in Merola et al. 1982 (no figures)

Cyanidium sulphuraria (Galdieri 1899) Ott in Ott & Seckbach 1994

Cells blue-green, spherical, 7–14 µm in diameter, forming powdery layer on stones. Autosporangium with 4–16(–32) trihedrally arranged autospores, 2–5 µm in diameter each.
Occurrence: Worldwide on stones in acid hot springs. Europe: Garderi in Italy.

2. Order **Goniotrichales** Skuja 1939

Thalli diffuse uniseriate filaments, sometimes arising from a pseudoparenchymatous base. Sexual reproduction unknown.

Important references: Starmach (1977), Bourrelly (1985), Vis & Sheath (1993), Kumano (2002)

Key to families:

1a Cells forming ± distinct filaments, each cell having one stellate chloroplast containing a central pyrenoid. Reproduction by monospores .... **1. Goniotrichaceae** (p. 23)

1b Thallus filamentous with false and true branching; cells comprise several parietal chloroplasts without pyrenoid. Reproduction by small spores (microgonidia) .... **2. Phragmonemataceae** (p. 23)

1. Family **Goniotrichaceae** G.M. Smith 1933

Cells located in ± distinct filaments; cells with stellate chloroplast containing a central pyrenoid. Reproduction by naked monospores.

Chroodactylon Hansgirg 1885

Asterocytis (Hansgirg 1885) Gobi et Schmitz in Engler et Prantl 1896

Unbranched or irregularly false branched filaments; cells elongated-elliptical to flat-spherical or rounded cubic with a thick gelatinous matrix. In alkaline waters containing *Cladophora* and *Rhizoclonium*.

Note: See Entwisle & Kraft (1984) and Vis & Sheath (1993) for taxonomic discussion on synonyms.

Chroodactylon ornatum (C. Agardh 1824) Basson 1979 (plate **3**: fig. **2**; plate **4**: figs. **3–4**)

Hormospora ramosa Thwaites in Harvey 1848, *Asterocytis ramosa* (Thwaites) Gobi 1879, *Asterocytis smaragdina* (Reinsch) Forti in De Toni 1907, *Chroodactylon ramosum* (Thwaites) Hansgirg 1885

Pseudofilaments with false branching, 0.024–10 mm long and surrounded by thick gelatinous matrix; cells rectangular to ellipsoidal with one blue, stellate chloroplast with central pyrenoid.Cell diameter 5.8–11.6(–20) µm, cell length 7.1–16.6 µm.
Occurrence: Epiphytic on aquatic plants *(Cladophora* and *Rhizoclonium)* or epilithic in hard-water ponds. Europe: Austria, Denmark, Hungary, The Netherlands, Norway, Poland, Portugal, Slovakia, Spain, Sweden, Switzerland and United Kingdom.

2. Family **Phragmonemataceae** Skuja 1939

Thallus filamentous with false and true branching; cells containing several parietal chloroplasts without pyrenoid. Reproduction by small spores (microgonidia).

Key to the genera:

1a Thalli and cells polymorphic; cells containing several parietal chloroplasts; colour grey-violet, pink or olive .... **1. Phragmonema**

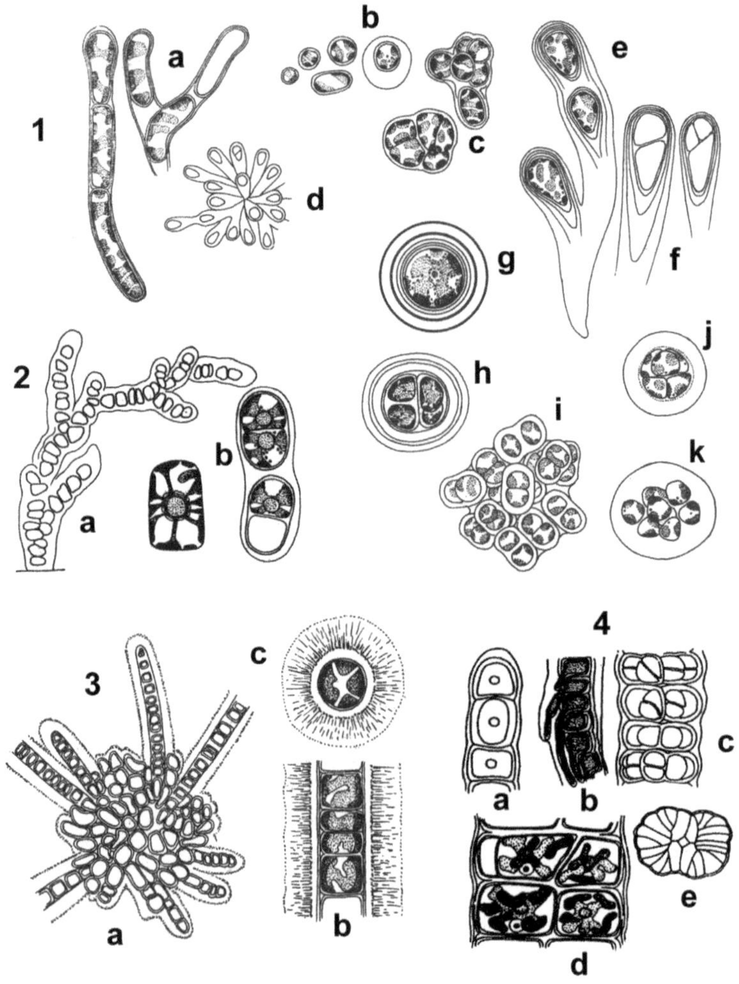

Plate 3: **Fig. 1.** *Phragmonema sordidum* (original author, Friedmann 1956): **a** – status *filamentosus*; **b** – status *coccoideus*; **c** – status *pseudoparenchymatosus*; **d–f** – status *dendroideus*; **g–i** – status *palmelloideus*; **j–k** – status *autosporogeneus*. **Fig. 2.** *Chroodactylon ornatum* (original authors, Geitler 1924, Belcher & Swale 1960): **a** – branching filament; **b** – cells with stellate chloroplast. **Fig. 3.** *Kyliniella latvica* (original author, Skuja 1926): **a** – discoid base and erect filaments; **b** – detail of erect filament; **c** – cross section of thread. **Fig. 4.** *Bangia atropurpurea* (original authors, Bourrelly 1970: a, d; Wołoszyńska 1935: b, c, e): **a** – an apex of filament; **b** – lower part of filament with developing rhizoids; **c** – multiseriate filament; **d** – filament with cells having stellate chloroplasts; **e** – cross-section of filament.

1b Filamentous arising from pseudoparenparenchymatous base; chloroplasts red-violet . . . . **2. Kyliniella**

1. Phragmonema Zopf 1882

Thalli and cells polymorphic; cells containing several parietal chloroplasts; colour grey-violet, pink or olive.

Phragmonema sordidum Zopf 1882 (plate **3**: fig. **1**)

6 different forms with different shape and arrangement of cells can be distinguished:

status *filamentosus*
status *coccoideus*
status *pseudoparenchymatosus*
status *dendroideus*
status *palmelloideus*
status *autosporogeneus*

Description given for the status *filamentosus*: filaments unbranched or false branched, containing only a few cells; cells cylindrical, 4–10 µm wide, 5–28 µm long with thin mucilage layer.
Occurrence: Austria (Vienna, Schönbrunn park), Germany (Berlin, botanical garden), Italy (cave near Neapel), Spain (Pico de Europa).

2. Kyliniella Skuja 1926

Filaments arising from a pseudoparenchymatous base; chloroplasts red-violet, without pyrenoid. Monospecific genus.

Kyliniella latvica Skuja 1926 (plate **3**: fig. **3**)

Pseudofilaments 2–3 cm long, growing from a discoid base with tightly arranged rectangular cells embedded in a gelatinous matrix. Cells 7.4–14.8 µm in diameter, 4.9–17.3 µm long; chloroplasts red-violet to blue.
Occurrence: Occurs epiphytic on *Phragmites*; Europe: Austria, Latvia (Lake Usma), Sweden, Baltic Sea (Kylin 1956).

3. Order **Compsopogonales** Skuja 1939

Uniseriate filaments with small cells making cortication around central axis row. Parietal chloroplast in young cells, later this chloroplast is fragmented into many small discoid chloroplasts. Asexual reproduction by monospores by an unequal division of the parental cell. Sexual reproduction unknown.

Important references: Starmach (1977); Vis et al. (1992); Rintoul et al. (1999); Kumano (2002); Yoon et al. (2006)

Note: See Rintoul et al. (1999) for taxonomic discussion.

1. Family **Compsopogonaceae** Schmitz in Engler et Prantl 1896

The only family in this order contains two genera, *Compsopogon* and *Compsopogonopsis*, which the latter does not occur in Europe (is found in East Asia and in Pacific Ocean islands).

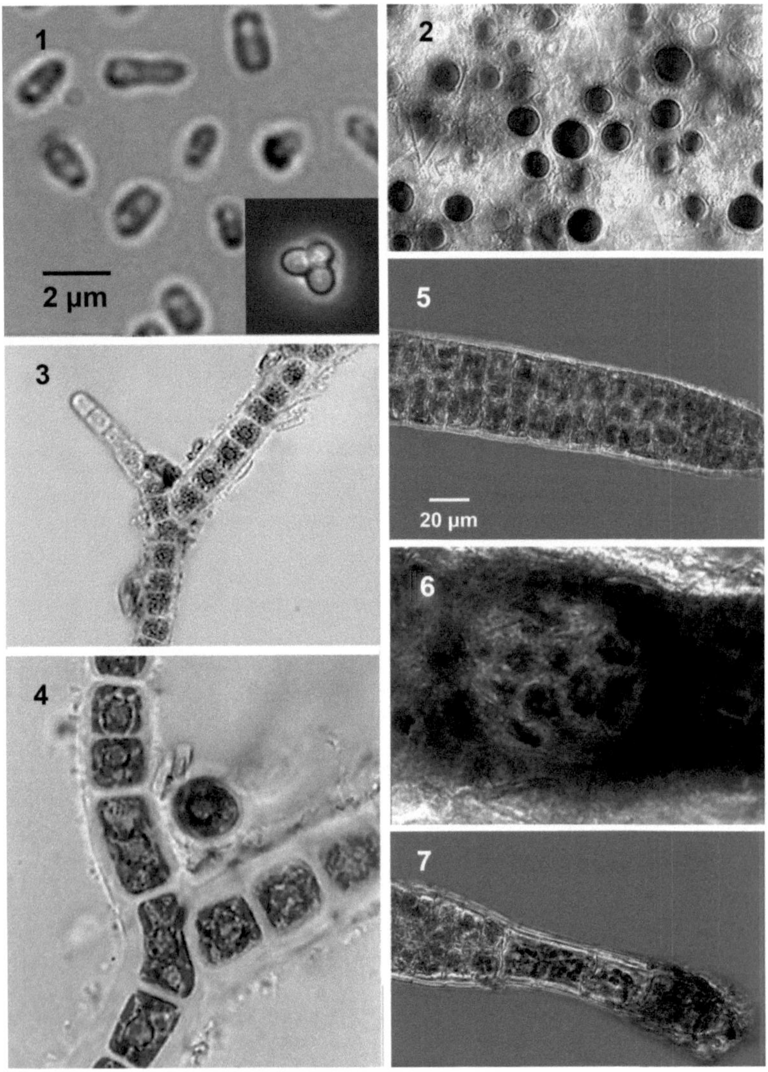

Plate 4: **Fig. 1.** *Cyanidioschyzon merolae* (photo: modified after G. Pinto 2007). **Fig. 2.** *Porphyridium purpureum*. **Fig. 3–4.** *Chroodactylon ornatum* (photo by F. Hindák in Wołowski, Kowalska & Hindák 2007) **Fig. 5–7.** *Bangia atropurpurea*: **5.** multiseriate thread; **6.** sporangium; **7.** basal part of thallus.

Compsopogon Montagne in Bory et Durieaux 1846

Thalli filamentous, branched, growing apically; composed of an axial thread covered with a cortex of compactly arranged smaller cells. Mainly tropical genus.

Key to the species:

1a Thalli up to 25 cm long **2**
1b Thalli 25–200–(300) cm long **3**
2a Thallus sparsely branched, branching from the base, branches directed forwards in an acute angle . . . . **1. C. chalybeus**
2b Branches loose, dispersed; older erect filaments have short, aculeate branchlets . . . . **2. C. aeruginosus**
3a Erect filaments sparsely branched . . . . **3. C. hookeri**
3b Erect filaments rich branched . . . . **4. C. coeruleus**

1. Compsopogon chalybeus Kützing 1849 (plate **5**: fig. **1**)

Compsopogon corinaldii (Meneghini 1841) Kützing 1857, *Lemanea corinaldii* Meneghini 1841

Thalli up to 20 cm high, branching from the base; threads at transversal walls of axial cells incised, in parts producing cortex 100–250 µm wide; cells of cortex 9–19 µm in diameter, 16–28 µm long; monospores 9–16 µm wide.
Occurrence: Tropical species growing in rivers and streams. Europe: Italia (tropical aquarium in Catania, Sicily), Ukraine.

2. Compsopogon aeruginosus (J. Agardh 1847) Kützing 1849 (plate **5**: fig. **2**; plate **6**: fig. **1–5**)

Pericystis aeruginosa J. Agardh 1847

Thalli heterotrichous, consisting of a prostrate holdfast and erect, dark blue-green filaments, 25–50 cm in length and 0.7 mm in diameter; prostrate system (holdfast) 180–390 µm in diameter; filaments rich branched with many short aculeate branchlets; cortex with 1–2 cell layers; cortical cells 12–30 µm in diameter, 16–40 µm long. Monosporangia spherical, 10–15 µm in diameter.
Occurrence: Occurs in streams; Europe: Austria (River Pulkau), Poland (in a botanical garden pond, Cracow).

3. Compsopogon hookeri Montagne 1846 (plate **5**: fig. **4**)

Thalli heterotrichous with basal disc, 140–250 µm in diameter and erect, dark blue-green filaments, up to 20–50 cm high, 0.5–2 mm in diameter, sparsely branched. Cortical layer with 1–2 cell layers; cortical cells 10–21 µm in diameter, 15–35 µm long. Monosporangia spherical, 13–19 µm in diameter; microsporangia in groups on main axis and branches, spherical 8–12 µm in diameter.
Occurrence: Known mainly from Asia; Europe: France (River Erth, affluent of Rhone).

4. Compsopogon coeruleus (Balbis ex C. Agardh 1824) Montagne 1846 (plate **5**: fig. **3**)

Conferva coerulea Balbis ex C. Agardh 1824; *Compsopogon oishii* Okamura 1915; *Compsopogon helwanii* El-Gamal et Salah El-Din 1999

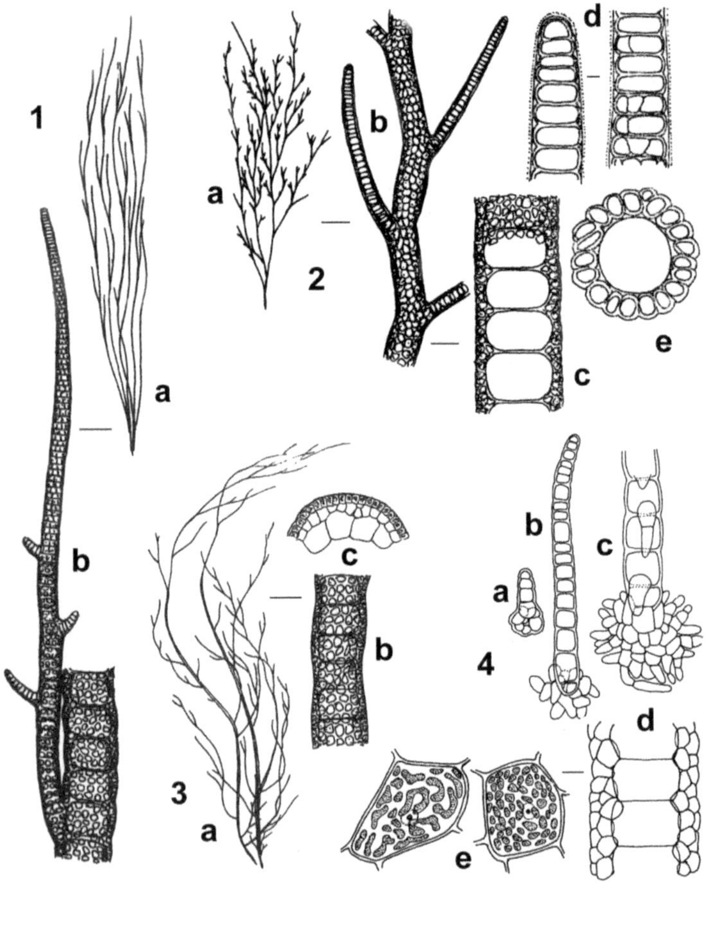

Plate 5: **Fig. 1.** *Compsopogon chalybeus* (original author, Kützing 1849): **a** – habitus of the plant; **b** – part of thallus with lateral branch. **Fig. 2.** *Compsopogon aeruginosus* (original author, Skuja 1938b): **a** – habitus of the plant; **b** – details of the thallus and branching; **c** – longitudinal section of thallus showing axial cells and cortical layer; **d** – longitudinal section of the young branch; **e** – cross-section of branch. **Fig. 3.** *Compsopogon coeruleus* (original authors, Jao 1941: a; Kützing 1849: b, c): **a** – habitus of the plant; **b** – detail of the thallus; **c** – part of cross-section of the branch. **Fig. 4.** *Compsopogon hookeri* (original author, Friedrich 1966): **a, b** – development stages of the young thallus; **c** – detail of a young branch; **d** – longitudinal section of an older branch, **e** – two cells with chloroplasts.

Thalli heterotrichous with basal disc, 300–500 µm in diameter; erect filaments dark blue-green, up to 10–80 cm high, 1–3 mm in diameter, rich and coarsely branched. Cortical layer with 2–3 cell layers; cortical cells 10–27 µm in diameter, 15–36 µm long. Monosporangia spherical, 15–20 µm in diameter.
Occurrence: In tropical rivers in Asia, Australia, Central and South America; Europe: Great Britain (Reddish Canal, Manchester), Malta, Spain.

4. Order **Bangiales** Schmitz in Engler 1892

For marine members, a life-cycle including sexual reproduction is known. Filaments pass through a diploid microphase called *Conchocelis* stage with pit connections (aperture in a cross wall between two cells with a plug and plug cap) *(Bangia)* and a macroscopic haploid phase without pit connections. In the genus *Boldia* (described only from America) thalli are saccate with 1–2 cells in thickness. Asexual reproduction by monospores; for freshwater species sexual reproduction is unknown.

Important references: Starmach (1977); Garbary et al. (1980); Sheath (1984); Müller et al. (1998, 2001); Kumano (2002)

Family **Bangiaceae** Engler 1892

Cells covered with a polysaccharide matrix divide transversally and longitudinally forming filaments with uniseriate to multiseriate structure. One species in freshwaters.

Note: According to Kumano (2002), the family Bangiaceae comprises two genera, *Bangia* and *Porphyra*. Nelson (2007) described a new monotypic genus *Bangiadulcis* (syn. *Bangia atropurpurea* (Roth) C. Agardh), which is distinguished from the other members of the Bangiales on the basis of cytological, reproductive and ecological characters.

Bangia Lyngbye 1819

Uni- or multiseriate filaments growing attached to stones or wood by rhizoids.

Bangia atropurpurea (Roth 1806) C. Agardh 1824 (plate **3**: fig. **4**; plate **4**: fig. **5**–**7**)

Conferva atropurpurea Roth 1806

Freshwater species; threads up to 15 cm long, unbranched, 13–98 µm in diameter, dark red, dark olive green or brownish. Young filaments consist of a single row of cylindrical cells, 6 µm wide and 8 µm long. Lateron, cells appear disc-shaped, 6–15 µm in diameter and 20 µm long. After longitudinal cell divisions, threads become polysiphonous (multiseriate). Cells 6 µm in diameter, 10 µm long are containing a stellate chloroplast with a central pyrenoid. Asexual reproduction by monospores, which develop in monosporangia located in mature filaments Sexual reproduction unknown.
Occurrence: In hard-water lakes and rivers, often nearby coastal areas; Europe: Found all over Europe.

Important references: Geesink (1973); Starmach (1977); Müller et al. (1998, 2001); Woolcot & King (1998); Kumano (2002)

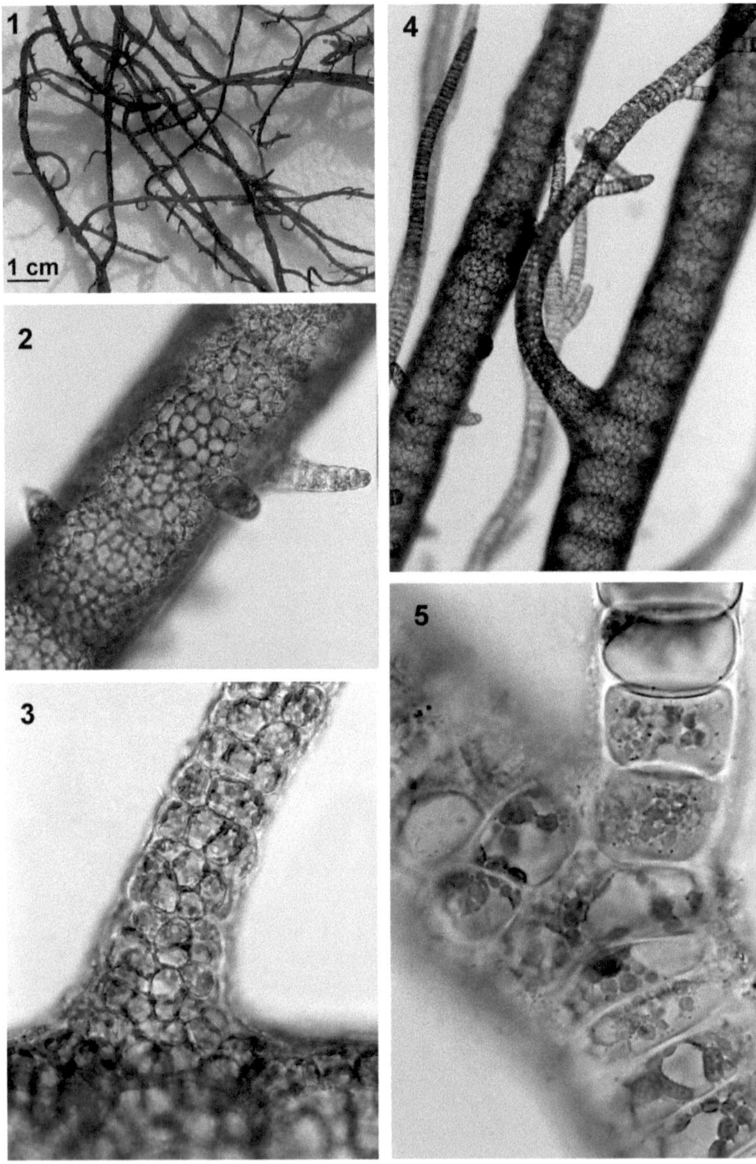

Plate 6: Fig. 1–5. *Compsopogon* cf. *aeruginosus* (photos: 1. Z. Žáková; 2–5. F. Hindák): **1.** habitus of the plant; **2.** part of thallus showing short spinous branchlets; **3.** multiseriate cortex cells of a side branch; **4.** plants with typical ramification; **5.** cells with parietal chloroplasts.

Note: Geesink (1973) has kept as synonyms *B. atropurpurea* and marine species *Bangia fuscopurpurea* (Dillwyn) Lyngbye 1819. According to Müller et al. (1998) all freshwater populations can be classified as a single species. See also note in the family description.

5. Order **Acrochaetiales** Feldmann 1953

Uniseriate, erect and branching filaments attached to the substratum by rhizoids. Asexual reproduction by monospores or by tetraspores; sexual reproduction monoecious or dioecious.

Important references: Israelson (1942); Starmach (1977); Garbary (1987); Necchi Jr. et al. (1993a, b); Harper & Saunders (2002); Kumano (2002)

1. Family **Acrochaetiaceae** Fritsch ex Taylor 1957

Only one genus known from European. For general generic concepts in the family Acrochaetiaceae see Kumano (2002).

1. **Audouinella** Bory 1823

Chantransia (De Candolle) Fries 1825; for other potential synonyms see Garbary (1987)

Thalli uniseriate, erect, branching filaments, with basal rhizoids. Chloroplasts violet red or grey-blue, laminate or ribbon-shaped; pyrenoids absent. Grows on stones, on water mosses or often on thallus of *Lemanea*. Monosporangia formed on both gametophytes and sporophytes. Carpospores terminal.

Note: Gametangia and carposporophytes are only rarely observed.

Important references: Skuja (1934); Starmach (1977); Necchi Jr. et al. (1993a, b); Necchi Jr. & Zucchi (1995); Zucchi & Necchi Jr. (2003).

Key to the species:

1a Thallus red or violet .... **1. A. hermannii**
1b Thallus grey-blue **2**
2a Lateral branches almost at an angle of 90° .... **2. A. serpens**
2b Lateral branches at acute angle **3**
3a Cells 1–3 times longer than wide .... **3. A. pygmaea**
3b Cells 3–7 times longer than wide **4**
4a Main filament ≥ 13 μm wide, monosporangia 12–18 μm wide, 18–24 μm long .... **4. A. leibleinii**
4b Main filament < 13 μm wide, monosporangia 8–11 μm wide, 9–12 μm long .. .. **5. A. chalybea**

Note: According to Skuja (1934) only red forms should be included into genus *Audouinella* and the blue forms are young chantransia-stages of the genus *Batrachospermum*. Also some later studies support this suggestion (Necchi Jr. & Zucchi 1997), however, in the actual key blue forms are still kept in the genus *Audouinella*.

1. Audouinella hermannii (Roth 1806) Duby in De Candolle 1830 (plate **7**: fig. **1**; plate **8**: **1–4**)

Chantransia hermannii (Roth) Desvaux 1809; *Chantransia violacea* Kützing 1845, *Audouinella miniata* Bory 1823, *Audouinella violacea* (Kützing) Hamel 1925, *Rhodochorton violaceum* (Kützing) Drew 1935

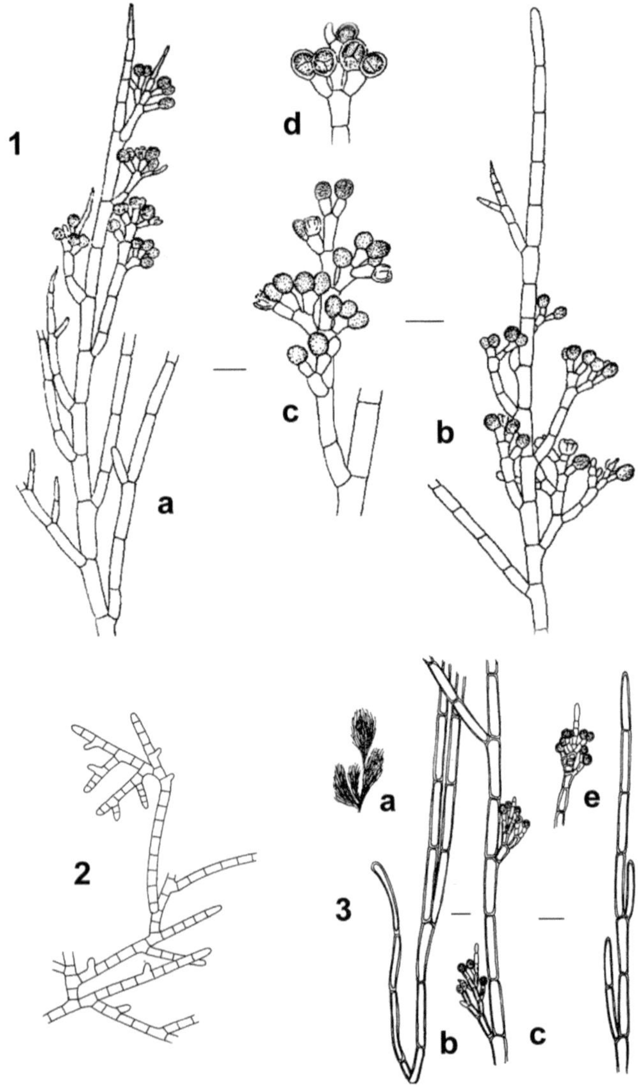

Plate 7: **Fig. 1.** *Audouinella hermannii*: (original author, Starmach 1977) **a–b** – parts of the plant with monosporangia; **c** – group of monosporangia; **d** – tetrasporangia. **Fig. 2.** *Audouinella serpens* (original author, Israelson 1942). **Fig. 3.** *Audouinella chalybea* (original author, Kützing 1845): **a** – habit of a plant; **b–e** – details of filaments with monospores.

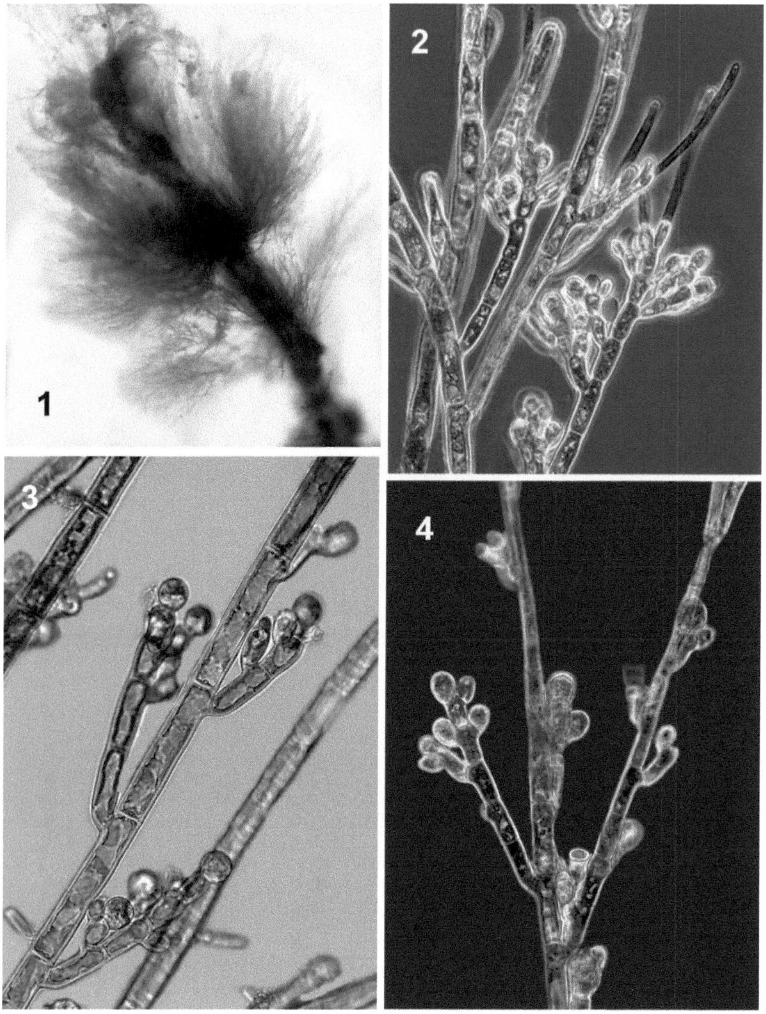

Plate 8: **Fig. 1–4.** *Audouinella hermannii*: **1.** tufts on thallus of *Lemanea*; **2.** parts of plant with branching filaments showing terminal monospores and hairs; **3–4.** filaments with monospores at the tip of short branches.

Plate 9: Fig. 1–6. *Audouinella chalybea*: **1.** dense tufts scraped off from stone; **2–3.** habitus of plant showing ramification; **4–5.** upper parts of plant showing monosporangia; **6.** typical filaments with lateral branches at narrow angle.

Thallus bright red to violet-red, forming small tufts, 3–15 mm high, attached to water mosses, stones or often on threads of *Lemanea*. Cells 7–16 µm wide, 2.5–5 times longer (35–67 µm); chloroplast ribbon-like; terminal hairs absent or abundant, varying in length. Reproduction through tetraspores and monospores. Monosporangia located at the tips of short lateral branches; ellipsoidal monospores (7–10 µm in diameter, 8–13(–16) µm long). Single tetrasporangia also terminating short lateral branches, ellipsoidal, 9–10 µm in diameter and 10–13 µm long. Polyoecious; 1 to 4 spermatangia terminating at small branches, ellipsoidal or cylindrical in shape, 3.5–4.0 µm in diameter, 4.5–5.5 µm long. Carpogonia sessile on terminating branchlets, bottle-shaped, 4–6 µm in diameter at basal part and 9.0–13.5 µm long. Gonimoblasts ± regularly globular, ca. 50–70 µm in diameter. Carposporangia ellipsoidal, inversely ovoid or pear-shaped, 9–11 µm in diameter and 10–15 µm long.
Occurrence: In running waters; this species seems to be tolerant for moderate organic pollution. Found in many countries all over Europe.

2. Audouinella serpens (Israelson 1942) Sheath 1984 (plate **7**: fig. **2**; plate **11**: fig. **5**)

Pseudochantransia serpens Israelson 1942

Thallus non-adherent, bluish-green, olive or dark green, straight filaments up to 5 mm with lateral branches protruding almost at an angle of 90°. Cells cylindrical, 16–18 µm in diameter, 30–60(–90) µm long; apex of terminal cells rounded. Terminal hairs absent. Chloroplasts parietal, ribbon-like, sometimes divided into segments. Monosporangia and propagation unknown.
Occurrence: found only on sediment in the littoral of Lake Våntjärn, central Sweden (Israelson 1942).

3. Audouinella pygmaea (Kützing 1843) Weber-van Bosse 1921 (plate **10**: fig. **2**; plate **11**: fig. **3–4**)

Pseudochantransia pygmaea (Kützing) Brand 1909

Note: Necchi Jr. et al. (1993b) synonymized *Audouinella leibleinii* Kützing with *A. pygmaea* according to morphological features. However, *A. chalybea* and *A. leibleinii* share more features (similar branching, similar cell length/width ratio >3) and therefore might be synonymous to *A. chalybea*. In this key, *A. pygmaea* is still kept as valid species.

Plants macroscopic, up to 20 mm in height, bluish, olive-green, sometimes brownish. Lateral branches in acute angle (<25°), almost attaching the main branch. Cells of main branch 5.5–25.0 µm in diameter, 14.0–66.5 µm long (length to width ratio is 1–3). Filaments attached to the substratum by a prostrate system formed from densely aggregated filaments. Monosporangia obovoidal or subspherical, 6–15 µm in diameter, 8–25 µm long. Carpogonia, tetrasporangia and carposporophytes not observed.
Occurrence: In running waters; Europe: Finland, Poland, Sweden, Switzerland, United Kingdom.

4. Audouinella leibleinii (Kützing 1845) Palmer in Hirsch et Palmer 1958 (plate **10**: fig. **1**)

Chantransia leibleinii Kützing 1845, *Pseudochantransia leibleinii* (Kützing) Israelson 1942, *Chantransia chalybea* (Roth) var. *leibleinii* (Kützing) Rabenhorst 1847

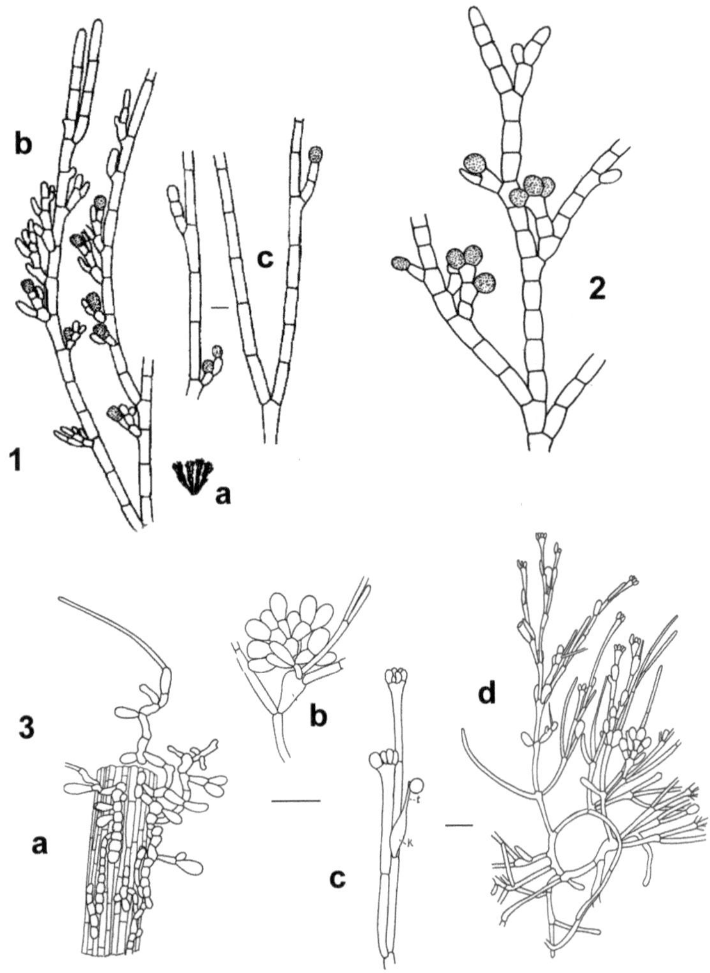

Plate 10: **Fig. 1.** *Audouinella leibleinii* (original authors, Kützing 1845: a, c; Israelson 1942: b): **a** – plant habitus; **b–c** – details of filaments showing branching and monospores. **Fig. 2.** *Audouinella pygmaea* (original author, Starmach 1977). **Fig. 3.** *Balbiania investiens* (original author, Sirodot 1876): **a** – part of a plant creeping along cortical filaments of *Batrachospermum*; **b** – carposporophyte with carposporangia; **c** – spermatangia at the top of branchlets and laterally standing carpogonium with spermatium at the top of a trichogyne; **d** – habit of the plant showing branching, spermatangia and young carposporophyte.

Plants tufted, 3–8 mm high, stiff, grey-green with bluish shade or brownish violet, when drying almost black. Branching abundant, irregular, lateral branches often reaching the heigth of main filaments making main filaments almost inconspicuous. Cylindrical cells uninuclear with a single parietal chloroplast, 13.0–18.5 µm in diameter, 45–85 µm long, 3–5 times longer than broad. Terminal cells with rounded apex, sometimes slightly swollen, terminal hairs absent. Monosporangia globular-ellipsoidal, 12–18 µm in diameter, 18–24 µm long; they develop on the short lateral branchlets, consisting of 1–4 cells.
Occurrence: In running waters on stones, sometimes on wood and mosses; Europe, rather rare species: recorded at least from Finland, Czech Republic, Great Britain, Sweden. Occurrence not clear because sometimes identified as *A. pygmaea* (see note).

5. Audouinella chalybea (Roth 1806) Bory 1823 (plate **7**: fig. **3**; plate **9**: fig. **1–6**)

Trentepohlia pulchella β *chalybea* C. Agardh 1824, *Trentepohlia pulchella aeruginosa* C. Agardh 1824, *Chantransia chalybea* (Lyngbye) Fries 1825, *Pseudochantransia chalybea* (Roth 1806) Brand 1909

Plants tufted, 2–5 mm high, grey-green with bluish shade, sometimes brownish. Branching abundant, lateral branches often reaching the height of main filaments making main filaments almost inconspicuous. Cells (6–) 8–12 (–13) µm in diameter, 30–80 µm long, 3–7 times longer than broad. Terminal cells with rounded apex. Chloroplast parietal, often segmented into spirally arranged ribbons. Terminal hairs mostly absent, rarely abundant. Monosporangia abundant or sparse, sometimes absent, 1–2 together terminating short lateral branchlets, globular-ellipsoidal, 8–11 µm in diameter, 9–13 µm long.
Occurrence: In springs and running waters; Europe: Belgium, Finland, Germany, Hungary, Poland, Sweden, Switzerland.

6. Order **Balbianiales** Sheath et Müller 1999

Monoecious uniseriate filaments, creeping among whorls of *Batrachospermum*; do not form rhizoids. Asexual reproduction by monospores. Spermatangia develop on top of elongated cells.

Important references: Israelson (1942); Starmach (1977); Garbary (1987); Sheath & Müller (1999); Kumano (2002)

Note: For revised description, taxonomic discussion and phylogeny see Garbary (1987) and Kumano (2002).

Family **Balbianiaceae** Sheath et Müller 1999

Characteristics like for the order. From Europe only one genus is known.

Balbiania Sirodot 1876

Thallus red purple; cells 10–15 times longer than wide; epiphytic on *Batrachospermum*. Monospecific genus.

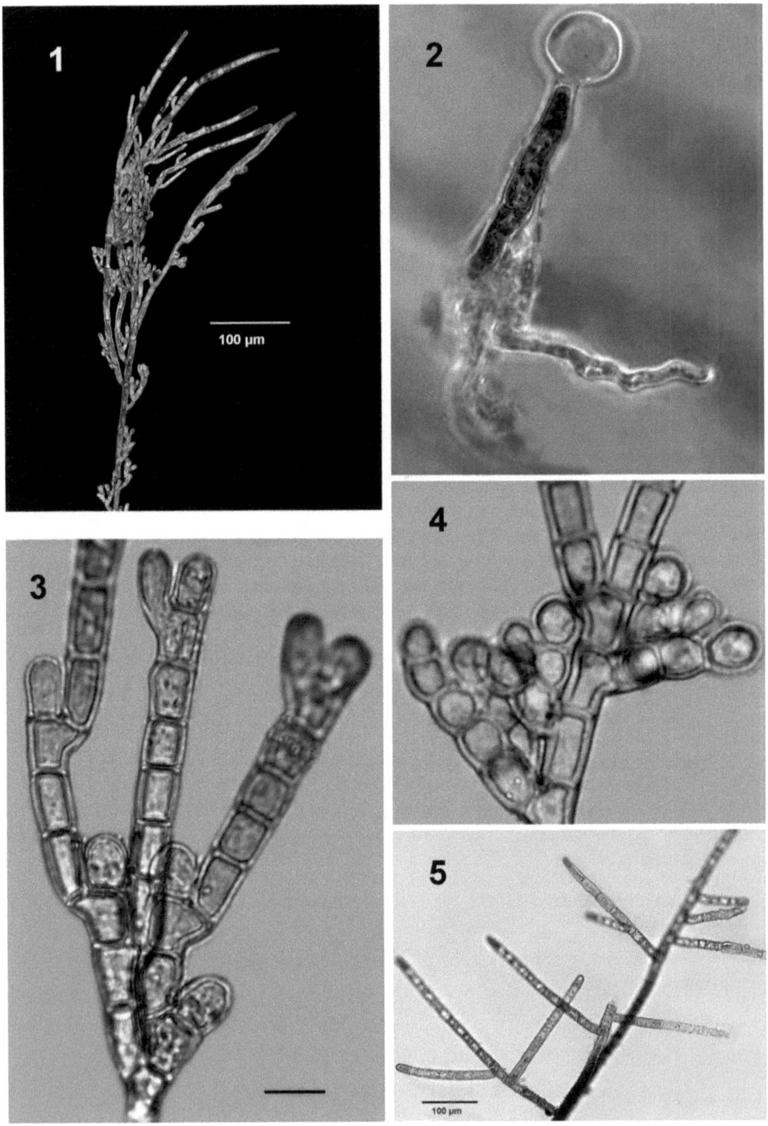

Plate 11: **Fig. 1–2.** *Audouinella chalybea*: **1.** plant showing ramification; **2.** germinating monospore with a short rhizoid; **Fig. 3–4.** *Audouinella pygmaea* (photo: J. Kostkevičienė): upper parts of plant showing monosporangia and typical short cells; **Fig. 5.** *Audouinella serpens* (photo: R. Bengtsson): plant with typical ramification.

Balbiania investiens (Lenormand ex Kützing 1849) Sirodot 1876 (plate **10**: fig. **3**)

Rhodocorton investiens (Lenormand 1849) Swale et Belcher 1963, *Batrachospermum rubrum* Hassall 1845, *Audouinella investiens* (Sirodot 1876) Garbary 1987

Plants purple-red; cells 10–15 times longer than their diameters. Terminal hairs commonly developed but often detached. Chloroplast ribbon-shaped, spirally. Monosporangia terminal on short lateral branches, spherical or ovoidal, 7–9 μm in diameter, 15–18 μm long. Spermatangia 2–5, forming a cluster at the tip of short lateral branches. Carpogonium on top of short lateral branches, ovoidal with a long slender trichogyne. After fertilization, carpogonium produces short gonimoblast filaments. Terminal cells of gonimoblast filaments produce single naked carpospores.
Occurrence in Europe: France, Germany, Ireland, Portugal, Sweden, United Kingdom.

7. Order **Batrachospermales** Pueschel et Cole 1982

Thalli uniaxial with irregular branchings; main axis either composed of cells forming on apices of whorls short branches or tubular branches or unbranched tufts with regular nodes. Asexual reproduction by monospores only in some species. For sexual reproduction, carpogonia with trichogynes are developed. The life cycle includes a macroscopic gametophyte phase and a small filamentous diploid chantransia-phase, which never produces tetraspores (meiosis takes place in terminal cells of the filament and the gametophyte is directly developing from that cell; '*Lemanea*'-type life-history; Dixon 1982).

Important references: Israelson (1942); Starmach (1977); Dixon (1982); Sheath (1984); Necchi Jr. & Entwisle (1990); Vis et al. (1998); Kumano (2002); Kapraun et al. (2007); Entwisle et al. (2009)

Note: According to Kumano (2002), this order includes three families: Batrachospermaceae, Psilosiphonaceae (does not occur in Europe) and Lemaneaceae. According to taxonomic revision by Entwisle et al. (2009) based on recent molecular studies the families Lemaneaceae and Psilosiphonaceae should be synonymized with the family Batrachospermaceae. In the actual key, the family Lemaneaceae is still kept as an own family.

Key to the families:

1a Plants with a main axis composed of an axial cell row and whorls of short branchlets; more or less mucilaginous, often with a bead-like habitus .... **1. Batrachospermaceae** (p. 39)

1b Plants tubular tufts with regularly placed swellings (nodes); carposporophytes (gonimoblasts) developed within the tubular thallus, spermatangia in form of papillae produced at nodes .... **2. Lemaneaceae** (p. 86)

1. Family **Batrachospermaceae** C. Agardh 1824

Plants uniaxial and irregularly branched. Central axis covered loosely with cortical filaments. Carpogonium-bearing branches arising either directly from periaxial cells and/or cells of fascicles or cortical filaments. Carpogonia symmetrical or

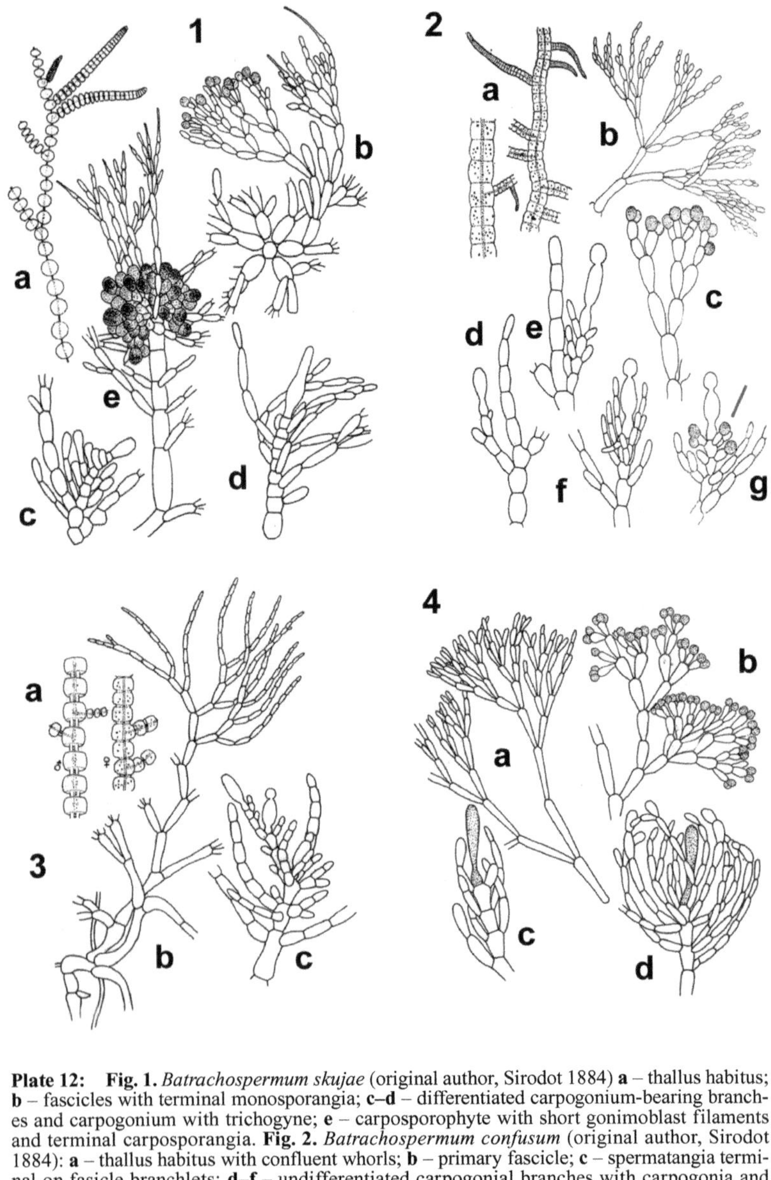

Plate 12: Fig. 1. *Batrachospermum skujae* (original author, Sirodot 1884) **a** – thallus habitus; **b** – fascicles with terminal monosporangia; **c–d** – differentiated carpogonium-bearing branches and carpogonium with trichogyne; **e** – carposporophyte with short gonimoblast filaments and terminal carposporangia. **Fig. 2.** *Batrachospermum confusum* (original author, Sirodot 1884): **a** – thallus habitus with confluent whorls; **b** – primary fascicle; **c** – spermatangia terminal on fasicle branchlets; **d–f** – undifferentiated carpogonial branches with carpogonia and short involucral filaments; **g** – spermatangia on involucral filaments of carpogonial branch. **Fig. 3.** *Batrachospermum boryanum* (original author, Sirodot 1884): **a** – main axis of male

asymmetrical with elongated conical or club-shaped trichogynes. Carposporophytes definite or indefinite in shape.

Note: According to the molecular studies by Entwisle et al. (2009), a new genus *Kumanoa* was proposed including all species from the *Batrachospermum* sections *Contorta* and *Hybrida* which is already considered in the actual key.

Key to the genera:

1a Central axis covered loosely with cortical filaments, which do not form a compact tissue **2**
1b Central axis covered with a compact cortex of pseudoparenchymatous type . . . . **4. Tuomeya**
2a Carposporophyte diffuse, gonimoblast filaments creeping along cortical filaments . . . . **3. Sirodotia**
2b Carposporophyte can easily be located **3**
3a Carposporophytes spherical or hemispherical, carpogonial branch more or less straight, not curved or spirally twisted . . . . **1. Batrachospermum**
3b Carposporophyte always central in the whorl and usually hemispherical, carpogonial branches curved, spirally coiled or twisted . . . . **2. Kumanoa**

1. **Batrachospermum** Roth 1797

Central axis loosely covered with cortical filaments, which do not form a compact tissue. Plants irregularly branched. Carposporophyte condensed in spherical or hemispherical formations. Carpogonium symmetrical or somewhat asymmetrical with elongated conical or clavate trichogyne. Gonimoblast filaments generally radially branched, sometimes appearing diffuse in the section *Turfosa*.
Batrachospermum is the most diverse genus among freshwater red algae and its classification is still under construction.

Note: According to their molecular studies, Entwisle et al. (2009) proposed to re-establish the section *Helminthoidea* Sirodot ex De Toni and include all species with irregular cortication and also *B. arcuatum*. In the following key the proposed change has not been applied because Entwisle et al. (2009) considered only some European taxa.

Key to the sections:

1a Carposporophyte sessile, hemispherical or aborted **2**
1b Carposporophyte pedicellate, spherical **4**
2a Carposporophyte consisting of both prostrate and erect gonimoblast filaments . . . . 3. Section **Turfosa**
2b Carposporophyte consisting of only erect gonimoblast filaments **3**
3a Whorls mostly reduced . . . . 2. Section **Setacea**
3b Whorls fully developed **4**
4a Carposporophyte large, 1 or 2, situated in the middle of whorls . . . . 4. Section **Virescentia**
4b Carposporophytes small, numerous, distributed in whorls . . . . 1. Section **Batrachospermum**

---

plant with separate whorls and female plant with somewhat confluent whorls; **b** – a primary fascicle with bulbous basal cells; **c** – carpogonia on undifferentiated carpogonium-bearing branch. **Fig. 4.** *Batrachospermum arcuatum* (original author, Kylin 1912): **a** – primary fascicle; **b** – fascicle with terminal spermatangia; **c–d** – carpogonium-bearing branches.

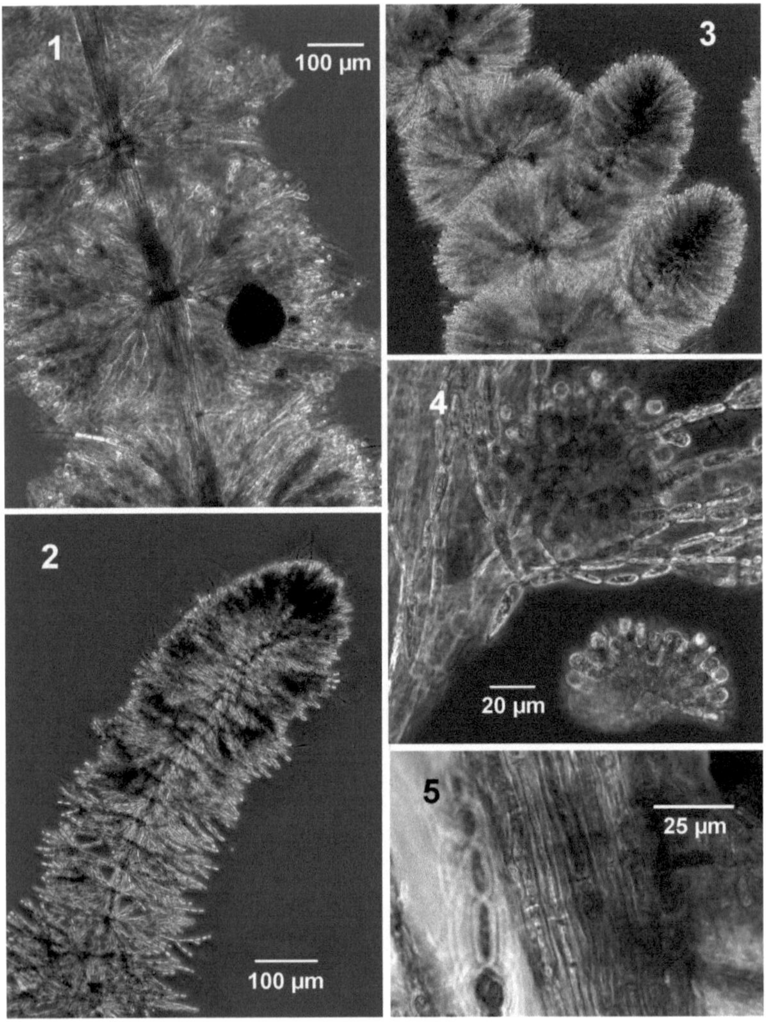

Plate 13: **Fig. 1–5.** *Batrachospermum skujae*: **1.** spherical whorl with a single carposporophyte; **2.** apex of young branch with dense confluent whorls; **3.** main axis with young branches; **4.** two carposporophytes showing carposporangia located terminally on gonimoblast filaments; **5.** main axis with regular cortical filaments.

1. Section **Batrachospermum**

Section *Moniliformia* Sirodot 1873, section *"Moniliformes"* Sirodot, section *Helminthoidea* Sirodot ex De Toni 1897, section *Carpocontorta* Sheath et al. 1986

Whorls well developed, carpogonial branches arising both from fascicle pericentral cells; carposporophytes spherical, rather small, pedicellate, numerous located in different parts of whorl.

Important references: Kylin (1912); Israelson (1942); Starmach (1984a); Necchi Jr. (1990); Vis et al. (1995, 1996a, b, 1998); Entwisle et al. (2009)

Note: Details of the taxonomic revision concerning the taxa belonging to section *Batrachospermum* can be found in Entwisle et al. (2009).

Key to the species of section *Batrachospermum*:

1a Monosporangia present, plants monoecious .... **1. B. skujae**
1b Monosporangia absent, plants monoecious or dioecious **2**
2a Spermatangia present on involucral (bract) filaments of the carpogonial branch .... **2. B. confusum**
2b Spermatangia absent on involucral filaments of the carpogonial branch ... **3**
3a Plants dioecious **4**
3b Plants monoecious **5**
4a Involucral filaments short, cortication irregular, carposporophytes scattered within the whorl .... **3. B. boryanum**
4b Involucral filaments long, cortication regular, carposporophytes peripheral .. .. **4. B. arcuatum**
5a Carposporophytes small within the whorl, involucral filaments long, cortication regular .... **5. B. gelatinosum**
5b Carposporophytes pedicellate, exserted to the surface of the whorl, cortication irregular **6**
6a Carpogonium ≤41 μm long, trichogyne ≤10 μm in diameter **6. B. anatinum**
6b Carpogonium ≥51 μm long, trichogyne ≥11 μm in diameter .. **7. B. fluitans**

1. Batrachospermum skujae Geitler 1944 emend. Vis et al. 1995 (plate **12**: fig. **1**; plate **13**: fig. **1–5**; plate **14**: fig. **1–5**)

Batrachospermum sporulans Sirodot 1884, nom. illeg., *B. moniliforme* var. *pulcherrimum* (Bory) Kützing

Plants monoecious, 1–6 cm high, more or less mucilaginous, grey-olive to grey-brown, richly branched; whorls confluent or distant, barrel-shaped, (400–) 700–900(–1000) μm in diameter with 1–6 carposporophytes scattered within the whorl; main axis with regular cortication (composed of cylindrical cells only); carpogonium 30–59 μm long with lanceolate to elongated lanceolate trichogyne, which is 7–13 μm in diameter; carpogonial branches 5–10 cells long, involucral filaments rather long; carposporophytes spherical, pedicellate, 50–152 μm in diameter, carposporangia obovoidal, 9–19 μm long, 8–14 μm in diameter; monosporangia present, 7–10 long, 7–10 μm in diameter.
Occurrence: In running waters. Known from rather many European countries from the North (Finland, Sweden) to the South Europe (Crimea, Portugal).

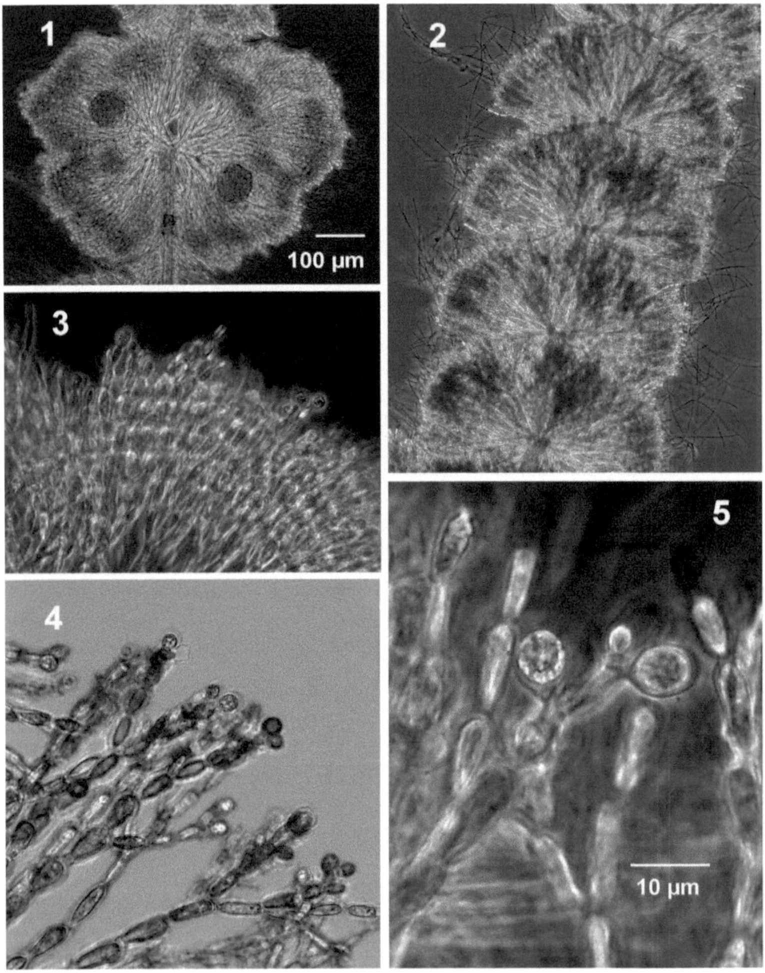

Plate 14: Fig. 1–5. *Batrachospermum skujae*: **1.** spherical whorl with two mature carposporophytes; **2.** main axis with confluent hemispherical whorls; **3–5.** fascicles with terminal monospores.

Plate 15: **Fig. 1–6.** *Batrachospermum confusum*: **1–2.** habitus of the plants with characteristic ramification and mostly confluent whorls; **3.** main axis with compressed whorls; **4–5.** carpogonium-bearing branches showing trichogynes and spermatangia at the tops of involucral filaments; **6.** groups of spermatangia terminal on the fascicle branches.

2. Batrachospermum confusum (Bory 1808) Hassall 1845 emend. Vis et al. 1995 (plate **12**: fig. **2**; plate **15**: fig. **1–5**)

Batrachospermum ludibundum var. *confusum* Bory 1808, *B. alpestre* Shuttleworth ex Hassall 1845, *B. setigerum* Rabenhorst 1859, *B. helminthosum* Sirodot 1884, nom. illeg. (non *B. helminthosum* Bory 1808), *B. crouanianum* Sirodot 1884, *B. distensum* Kylin 1912, *B. boryanum* var. *distensum* (Kylin) Israelson 1942, *B. fruticulosum* Drew 1946, *B. fernandesianum* Reis 1954, *B. confusum* f. *spermatogloberatum* Reis 1962

Note: For synonyms see also Vis et al. (1995) and Necchi Jr. (1990). – Based on the identical sequence data from the *rcb*L gene, Stewart & Vis (2007) proposed *B. confusum* to be included into *B. anatinum* as forma *B. anatinum* f. *confusum* (Bory) Stewart et Vis.

Plants monoecious, 3–10 cm high, very mucilaginous, grey with olive-yellowish or brownish shade; cells in the lower parts of fascicles are 10 times longer than wide; main axis with inflated irregular cortication; whorls elliptical or bulgy, 412–1000 (–1600) µm wide, coalescent, sometimes slightly separated; carpogonial branches 4–9 cells long, with short involucral filaments having apical spermatangia; carpogonium 14–34 µm long with clavate or ellipsoidal trichogyne; carposporophytes 1–19 scattered within the whorl, spherical, pedicellate, 56–100–130 µm in diameter, mostly in the outer part of whorl; carposporangia obovoidal, 6–11 × 8–14 µm.
Occurrence: In running waters; all over Europe.

3. Batrachospermum boryanum Sirodot 1874 emend. Vis et al. 1995 (plate **12**: fig. **3**; plate **16**: fig. **1–6**; plate **17**: fig. **1–6**)

Batrachospermum ectocarpoideum Skuja ex Flint 1949

Plants dioecious (according to Starmach 1977 polyoecious, bisexual specimens being more frequent), thallus 1–8 (–10) cm high, very mucilaginous; cells in the lower part of fascicles 4–5 times longer than wide, cell walls thick; whorls separated or confluent, spherical (337–)600–1100(–1200) µm in diameter; lower cells of fascicles swollen at the ends, 8–14 (–20) µm wide, 20–80(–100) µm long; carpogonial branches 4–8 cells long; carpogonium 15–35 µm long with inflated clavate trichogyne, 6–10 µm in diameter with short involucral filaments consisting of barrel-shaped cells; 1–30 spherical, pedicellate carposporophytes scattered within the whorl, (68–)120–160 µm in diameter; carposporangia obovoidal, 7–14 × 8–17 µm.
Occurrence: In springs and streams all over Europe.

4. Batrachospermum arcuatum Kylin 1912 emend. Vis et al. 1995 (plate **12**: fig. **4**; plate **18**: **1–4**)

Plants dioecious (polyoecious?), 5–10 cm high, very mucilaginous, grey-green, richly branched; whorls separated or confluent, barrel-shaped, 428–727 µm in diameter; main axis with regular cortication consisting of cylindrical cells only; fascicles form bush-like branches in apical parts; spermatangia on top of fascicles; carpogonial branches 5–9 cells long; carpogonium 28–39 µm with clavate trichogyne 6–12 µm in diameter surrounded by long involucral filaments which are bent inwards; carposporophytes 1–3, peripheral, spherical, pedicellate, 92–129 µm in diameter; carposporangium obovoidal, 10–15 µm long, 7–12 µm in diameter.
Occurrence: In running waters. Europe: Belgium, British Isles, Crimea, Finland, Lithuania, Poland, Portugal, Sweden.

Plate 16: **Fig. 1–6.** *Batrachospermum boryanum*: **1–2.** habitus of the plant; **3.** fascicle showing barrel-shaped cells; **4.** apex of branch with dense whorls; **5.** main axis with spherical whorls; **6.** regular cortication of the main axis.

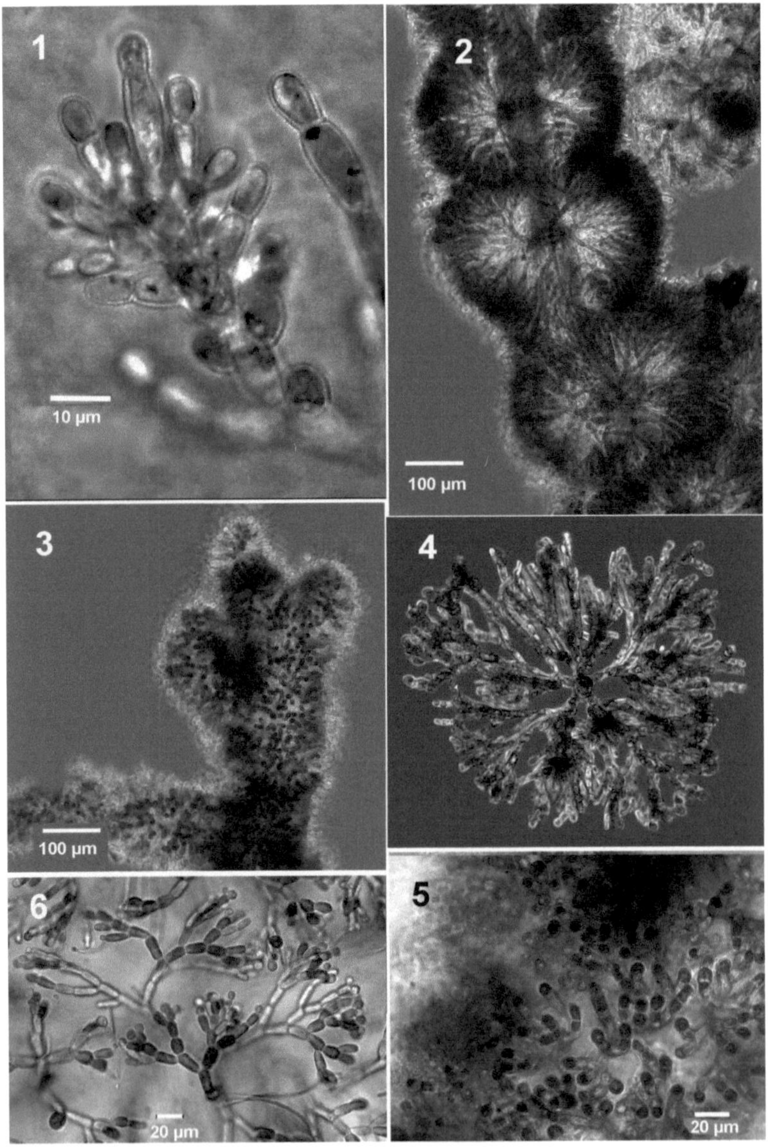

Plate 17: Fig. 1–6. *Batrachospermum boryanum*: **1.** carpogonium-bearing branch; **2.** main axis with spherical whorls; **3.** part of thallus showing young branches; **4.** primary fascicle; **5–6.** fascicles with barrel-shaped cells.

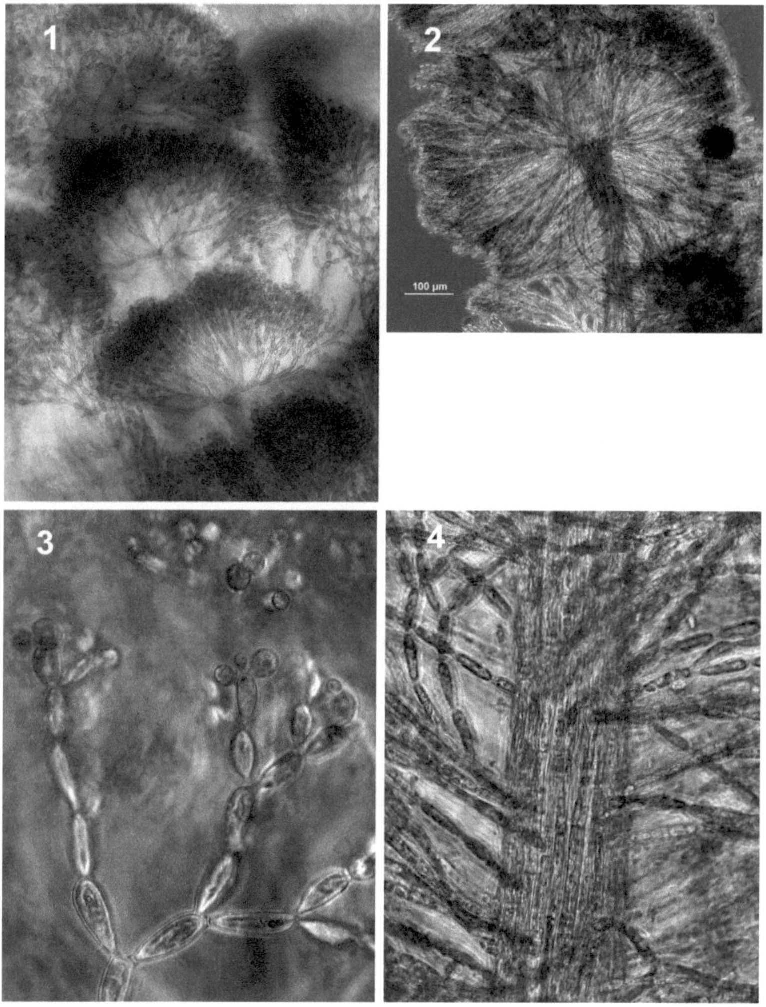

Plate 18: **Fig. 1–4.** *Batrachospermum arcuatum*: **1.** main axis with confluent whorls; **2.** main axis showing spherical, compressed whorls with a single carposporophyte; **3.** fascicles with terminal spermatangia; **4.** main axis with regular cortication.

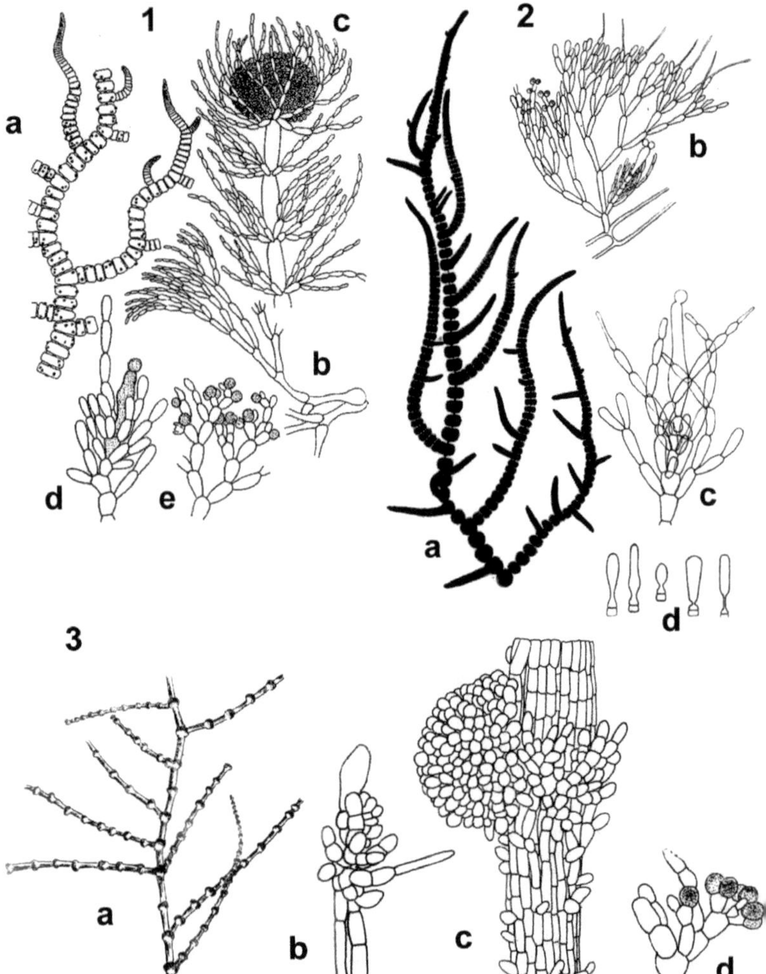

Plate 19: **Fig. 1.** *Batrachospermum anatinum* (original author, Sirodot 1884): **a** – general view of the thallus with confluent whorls; **b** – primary fascicle with swollen basal cell; **c** – stalked carposporophyte with dense situated carposporangia; **d** – undifferentiated carpogonium-bearing branch and carpogonium with lanceolate trichogyne; **e** – spermatangia on top of fascicles. **Fig. 2.** *Batrachospermum gelatinosum* (original authors Sirodot 1884 and Kylin 1912): **a** – general view of thallus with separate and confluent whorls; **b** – primary fascicle with terminal spermatangia and differentiated carpogonium-bearing branch; **c** – carpogonium with relatively long bracts and spermatium at the top of trichogyne; **d** – different types of trichogy-

5. Batrachospermum gelatinosum (Linnaeus 1753) De Candolle 1801 emend. Vis et al. 1995 (plate **19**: fig. **2**; plate **20**: fig. **1–6**; plate **21**: fig. **1–6**; plate **22**: fig. **1–6**)

Batrachospermum ludibundum var. *pulcherrimum* Bory 1808, B. *ludibundum* var. *caerulescens* Bory 1808, *B. ludibundum* var. *stagnale* Bory 1808, *B. hybridum* Bory 1823, *B. durum* C.A. Agardh 1824, *B. moniliforme* Sirodot 1884, *B. decaisneanum* Sirodot 1884, *B. radians* Sirodot, *B. reginense* Sirodot 1884, *B. corbula* Sirodot 1884, *B. densum* Sirodot 1884, *B. pygmaeum* Sirodot 1884, *B. pyramidale* Sirodot 1884, nom. illeg., *B. godronianum* Sirodot 1884, *B. moniliforme* var. *isoeticola* Skuja 1928, *B. corbula* var. *alcoense* Reis 1954, *B. arcuatoideum* Reis 1973

Plants monoecious, 2–20 cm high, more or less mucilaginous; whorls distant to confluent, sometimes pressed, spherical or barrel-shaped, (200–)500–800(–1100) µm in diameter, whorl branches often terminating in hairs; main axis with regular cortication; spermatangia on top of fascicles; carpogonial branches 3–10 cells long; carpogonium 20–68 µm long with clavate to lanceolate trichogyne, 5–17 µm in diameter, involucral filaments long and numerous; carposporophytes spherical, pedicellate, (40–)70–140 µm in diameter, 1–11 per whorl within the whorls at various distances from the main axis, carposporangium obovoidal, 6–12 × 8–16 µm.
Occurrence: Common in running waters all over Europe.

Note: This species is very variable and some local forms have been described based on different thallus morphology, e.g., fo. *pyramidale* Israelson 1942 (syn. *B. pyramidale* Sirodot, *B. pygmaeum* Sirodot, *B. moniliforme* var. *filamentosum* Rabenhorst, *B. vagum* var. *coerulescens* Rabenhorst), fo. *densum* (Sirodot) Israelson (syn. *B. densum* Sirodot) (Starmach 1984a).

6. Batrachospermum anatinum Sirodot 1884 emend. Vis et al. 1995 (plate **19**: fig. **1**; plate **23**: fig. **1–6**; plate **24**: fig. **1–5**)

Batrachospermum ectocarpum Sirodot 1884, *B. pulchrum* Sirodot 1884, *B. stagnale* (Bory) Hassall 1845, *B. moniliforme* var. *stagnale* (Bory) C. Agardh 1824

Note: Vis et al. (1995) and Vis & Sheath (1996) consider *B. pulchrum* as a species.

Plants monoecious (according to Starmach 1977 polyecious), thallus 2–12 cm high, very mucilaginous, grey or brown-olive, sometimes violet; cortication irregular, whorls confluent or separated, elliptical or bulgy, less frequently spherical, (479–)700–1500(–1900) µm wide; primary fascicles 12–20 cells storeys, distinctly bent at the apical parts, lower cells are swollen at the tops, 8–18 µm wide, 25–80 µm long; carpogonial branches 3–8 cells, carpogonium 17–41 µm long with inflated clavate or lanceolate trichogyne, 6–10 µm in diameter and with long numerous involucral filaments; carposporophytes 1–11 scattered within and exerted from the whorl, spherical, pedicellate, (69–)130–185 µm in diameter, carposporangia obovoidal, 8–16 µm long, 7–14 µm in diameter.
Occurrence: In creeks and streams all over Europe.

---

nes. **Fig. 3.** *Batrachospermum atrum* (original author, Sirodot 1884): **a** – habitus of plant; **b** – differentiated short carpogonium-bearing branch; **c** – hemispherical carposporophyte; **d** – spermatangia on top of fascicles.

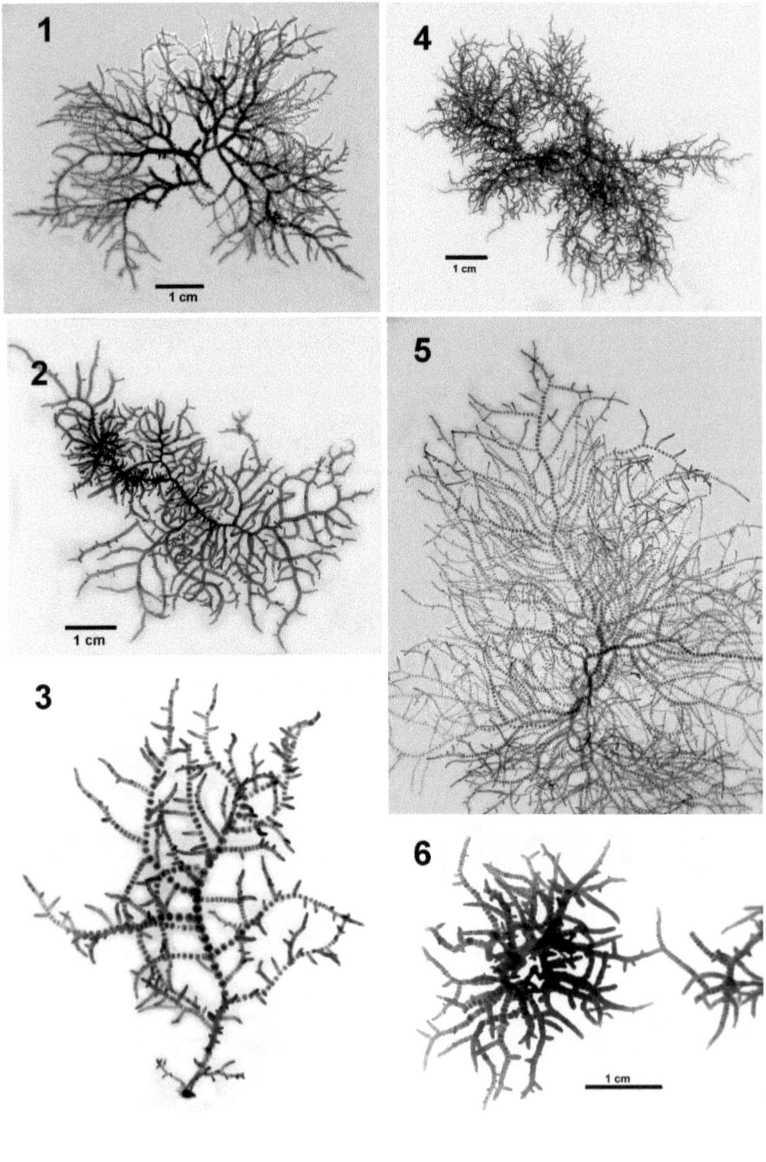

Plate 20: Fig. 1–6. *Batrachospermum gelatinosum*: habitus types of plants showing different types of structure and ramification.

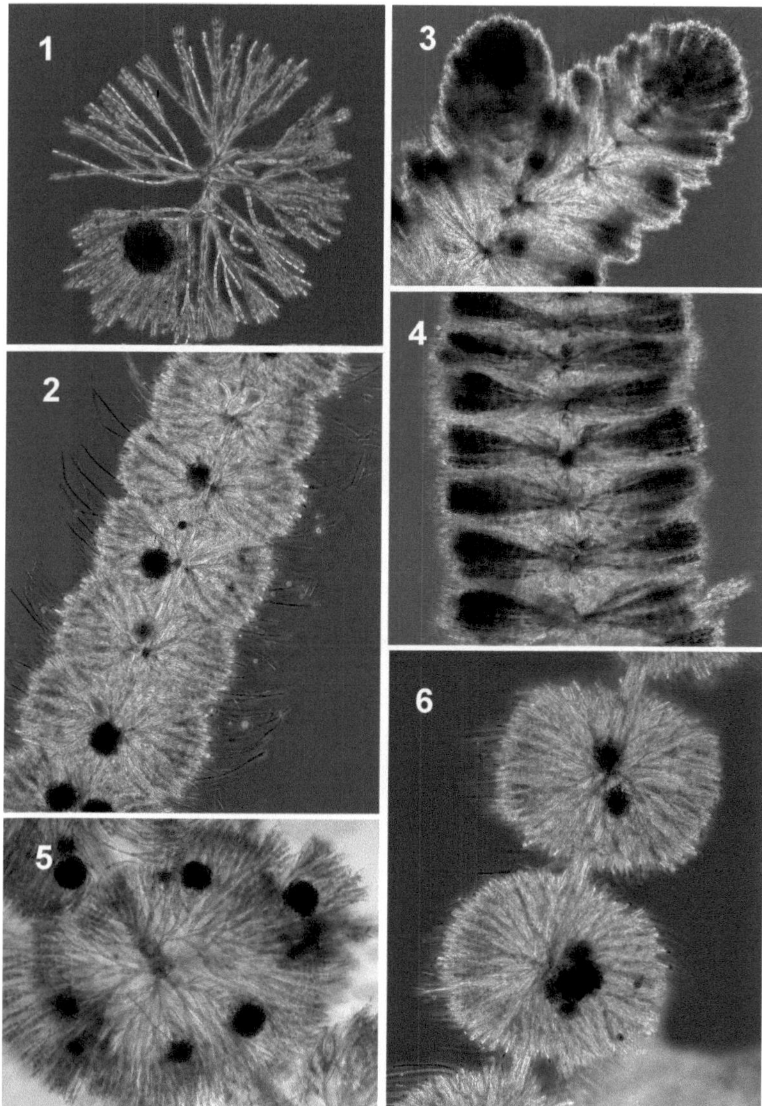

Plate 21: **Fig. 1–6.** *Batrachospermum gelatinosum*: **1.** primary fascicle with one carposporophyte; **2.** main axis with compressed whorls having some carposporophytes and fascicles with terminal hairs; **3.** upper part of the thallus with young branches; **4.** dense compressed whorls; **5–6.** main axis with several carposporophytes within the spherical whorls.

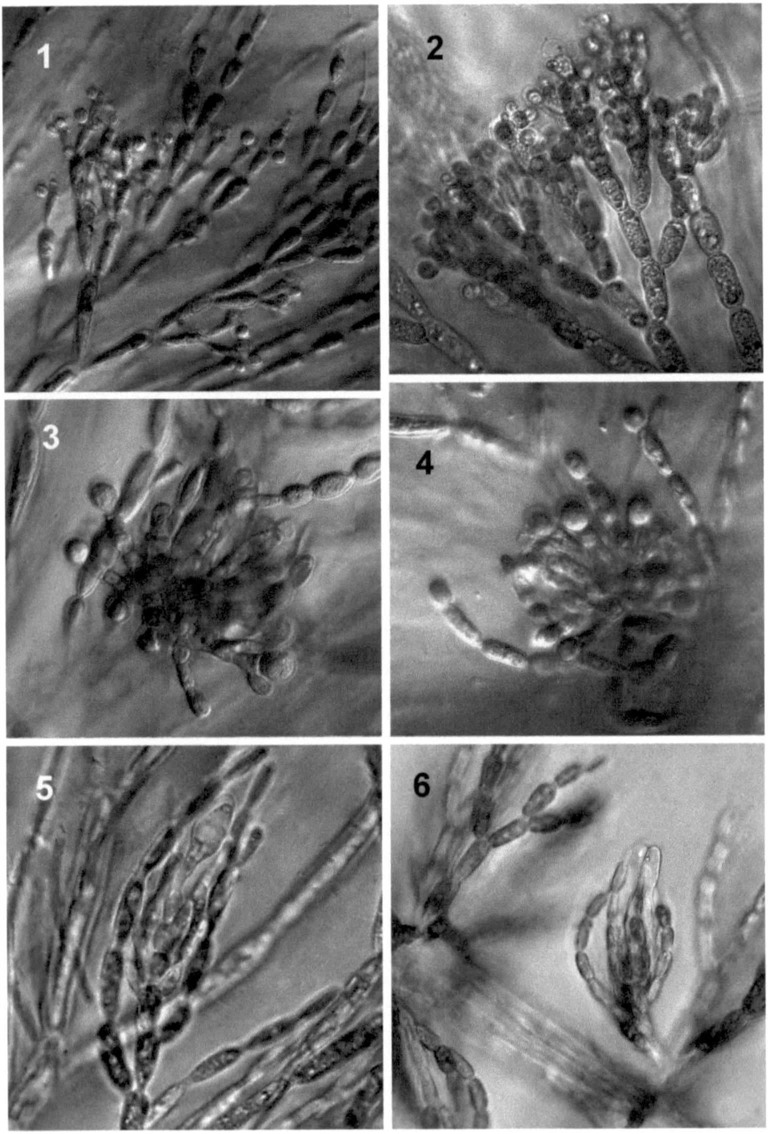

Plate 22: **Fig. 1–6.** *Batrachospermum gelatinosum*: **1–2.** groups of spermatangia terminal on fascicles; **3–4.** carposporophytes with carposporangia at the top of gonimoblast filaments; **5–6.** carpogonium-bearing branches with carpogonium and its trichogyne.

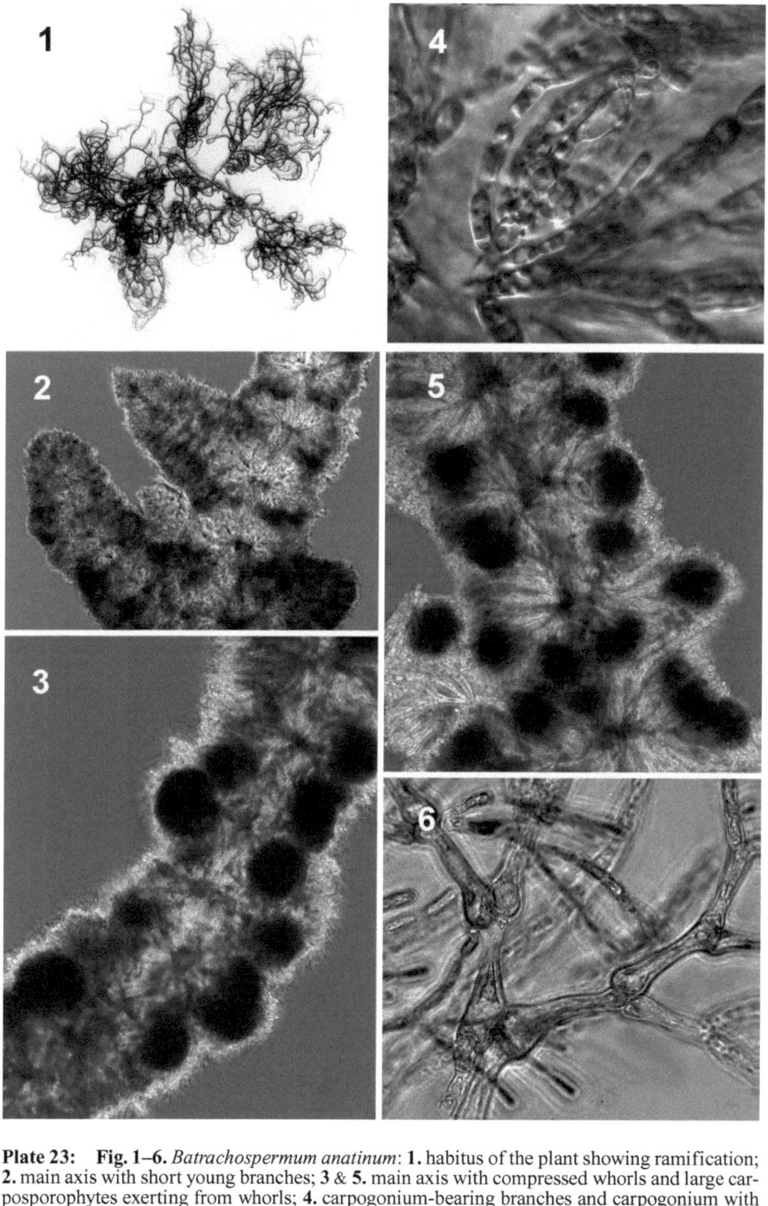

Plate 23: **Fig. 1–6.** *Batrachospermum anatinum*: **1.** habitus of the plant showing ramification; **2.** main axis with short young branches; **3** & **5.** main axis with compressed whorls and large carposporophytes exerting from whorls; **4.** carpogonium-bearing branches and carpogonium with a lanceolate trichogyne; **6.** swollen basal cells of primary fascicles.

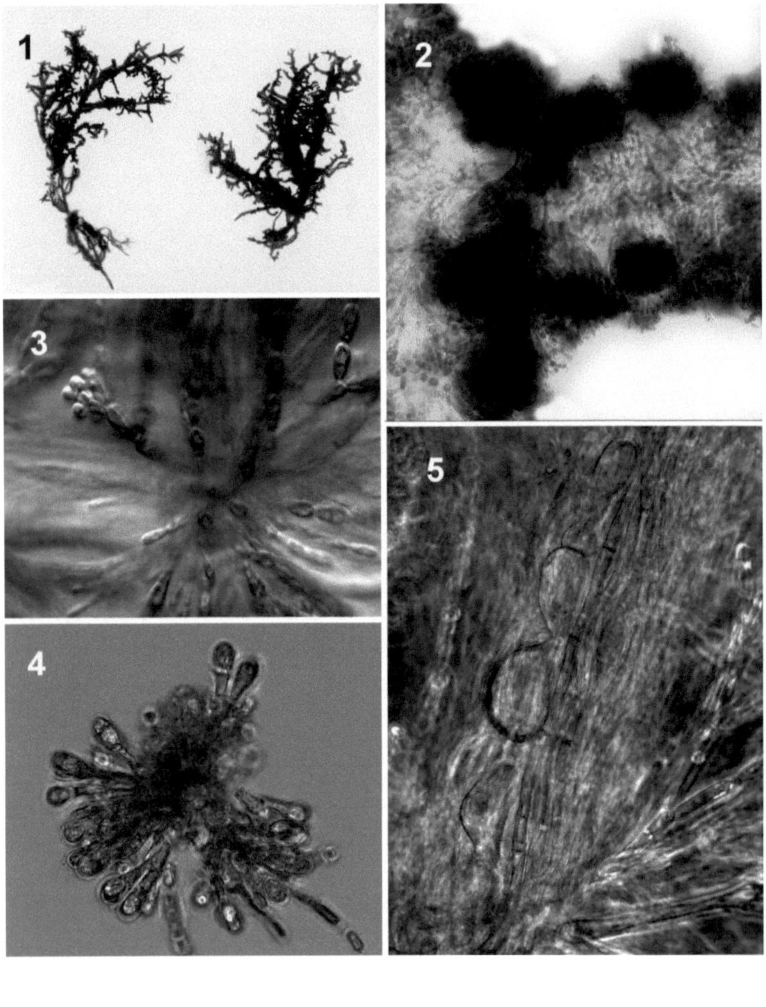

Plate 24: Fig. 1–5. *Batrachospermum anatinum*: **1.** plant habitus; **2.** main axis with compressed whorls and several large carposporophytes exerting from whorls; **3.** spermatangia terminal on fascicles; **4.** carposporophyte showing carposporangia terminal on gonimoblast filaments; **5.** irregular cortication of the main axis with inflated and cylindrical cells.

7. Batrachospermum fluitans Kerner 1882 emend. Vis et al. 1995 (no figure)

Plants monoecious, main axis with inflated irregular cortication; whorls confluent, spherical, 466–664 µm in diameter; carpogonial branches 4–6 cells long, carpogonium 51–65 µm long, 11–15 µm in diameter with lanceolate trichogyne; carposporophytes 1–30 per whorl, pedicellate, spherical, 64–93 µm in diameter.
Occurrence: Known only from Austria, near Mühlau, Innsbruck.

2. Section **Setacea** De Toni 1897

Section *Moniliformia* subsection *Capillacea* Sirodot 1873, section *Moniliformia* subsection *Setacea* Sirodot 1875, section *'Setaces'* Sirodot 1884

Fascicles very short and reduced; carpogonial branches very short and arising from periaxial cells, carpogonia with clavate or urn-shaped trichogyne; carposporophytes on main axis as wart-like protuberances.
From this section, only one species has been recorded for Europe.

Important references: Starmach (1977); Necchi Jr. (1990); Necchi Jr. & Entwisle (1990); Sheath et al. (1993a)

Batrachospermum atrum (Hudson 1898) Harvey 1841 (plate **19**: fig. **3**; plate **25**: fig. **1–6**)

Conferva atra Hudson 1798, *Batrachospermum tenuissimum* Bory 1823, *B. dillenii* Sirodot 1884, *B. gallaei* Sirodot 1884, *B. angolense* W. West et G.S. West 1897, *Sirodotia angolensis* (W. West et G.S. West 1897) Skuja in Reis 1960

Note: Sheath et al. (1993a) suggested to treat *B. atrum* as a synonym to *B. orthostichum* Skuja 1931 (not known from Europe), which according to Necchi Jr. is located in the section *Turfosa* (Necchi Jr. 1990, Kumano 2002).

Plants polyoecious, thallus 2–5(–13) cm high, poorly mucilaginous, grass-green or olive; whorls reduced, ovoid to conical, separated (in older plants compressed) only 100–240 µm wide, fascicles consist of 3–6 cell storeys, secondary fascicles numerous; spermatangia spherical to obovoidal, 4–6 µm in diameter, terminally located on fascicles and involucral filaments; carpogonial branches arising from periaxial cells and fascicles, consisting of 3–5 cells, slightly curved, with short involucral filaments; carpogonium 4–7 µm in diameter at the apex, 15–27 µm long, trichogyne inflated-ovoid, urn-shaped or ellipsoidal; carposporophytes single or paired, hemispherical, protruding wart-like from whorls, 80–170 µm wide, 50–100 µm high, carposporangium obovoidal, 6–8 µm wide, 7–12 µm long.
Occurrence: In standing and running waters all over Europe.

3. Section **Turfosa** Sirodot 1873 emend. Necchi 1990

Section *Turfosa* Sirodot 1873, section *Turficola* De Toni 1897, section *'Turficoles'* Sirodot 1884

Carpogonial branches straight, short, arising from periaxial cells of fascicles; carpogonia sessile or indistinctly stalked; carposporophytes central, single or in pairs; carpogonial filaments of two types, which is erect and prostrate.

Important references: Starmach (1977); Necchi Jr. (1990); Kumano (1993, 2002); Sheath et al. (1994c); Müller et al. (1997)

Plate 25: **Fig. 1–6.** *Batrachospermum atrum*: **1.** tuft of the plant; **2.** thallus showing reduced whorls and ramification; **3.** apices of young branches; **4.** main axis with reduced whorls, apex of a new branch and carpogonium with spermatium at tip of the trichogyne; **5.** main axis and a whorl with two carpogonia; **6.** main axis with cortical filaments.

Note: According to Starmach (1977), *B. globosporum* Israelson and *B. vogesiacum* Schultz ex Skuja belong to section *Turfosa* (Turficola). Recently *B. globosporum* was included into the new genus *Kumanoa* (Entwisle et al. 2009). Sheath et al. (1994c) maintained *B. vogesiacum* in section *Turfosa*, but Necchi Jr. (1990; as *B. flagelliforme*) and Kumano (2002) shifted it into the section *Virescentia*. In the actual key, *B. vogesiacum* is kept in section *Virescentia*.

Key to the species:

1a Whorls reduced, primary fascicles (2–)3–7(–8) cell storeys, monosporangia absent . . . . **1. B. keratophytum**

1b Whorls well developed, primary fascicles 8–15(–17) cell storeys, monosporangia present . . . . **2. B. turfosum**

1. Batrachospermum keratophytum Bory 1808 emend. Sheath, Vis et Cole 1994 (plate **26**: fig. **1–5**)

Batrachospermum vagum (Roth) C. Agardh var. *keratophytum* (Bory 1808) Sirodot 1884, *B. suevorum* Kützing 1849

Plants monoecious, 3.0–7.5 cm in height, bluish green, poorly mucilaginous, rigid, 160–500 µm in diameter; branching sparse, pseudodichotomous, main branch indistinct, apices straight; whorls poorly defined, obconical or compressed and indistinct, contiguous; cortical filaments well developed; primary fascicles straight, (3–) 4–7(–8) cell storeys; 1–2 hairs arising from each terminal cell, variable in length, base inflated; secondary fascicles numerous, reaching the length of primary fascicles; spermatangia spherical, terminal or subterminal on secondary fascicles, rarely on primary fascicles, 6–9 µm in diameter; carpogonial branches straight, on pericentral, rarely on proximal cells, short, 17–30 µm long; involucral filaments short, numerous; carpogonium symmetric, 36–47 µm long, with long clavate trichogyne; carposporophyte sessile, 1 per whorl, semi-spherical, larger than the radius of the whorl, 100–250 µm high, 55–130 µm wide; gonimoblast filaments of 2 types: (1) radially branched, erect filaments composed of ellipsoidal or barrel-shaped cells, 5–12 x 9–18 µm; (2) diffuse prostrate composed of cylindrical cells, 4–7 µm in diameter, 15–25 µm long. Carposporangia subspherical or spherical, 9–12 × 10–13 µm.
Occurrence: In standing and running waters all over Europe.

2. Batrachospermum turfosum Bory 1808 emend. Sheath, Vis et Cole 1994 (plate **27**: fig. **1**; plate **28**: **1–6**; plate **29**: **1–6**; plate **30**: **1–6**)

Chara gelatinosa var. *vaga* Roth 1797, *Batrachospermum vagum* (Roth) C. Agardh 1812, *B. moniliforme* var. *vagum* Roth 1800, *B. turfosum* var. *undulato-pedicellatum* Kumano et Watanabe 1983

Plants monoecious, 2–15(–20) cm high, bluish or deep green, moderately mucilaginous, bushy, 300–1000 µm wide, main branch indistinct; whorls already well developed in young branches, dense, obconical, barrel-shaped or compressed and indistinct, contiguous in old branches; primary fascicles have 8–15(–17) cell layers of proximal cells, pear-shaped, fusiform or egg-shaped; hairs numerous, short with inflated base; secondary fascicles numerous, along the whole internode area, reaching the length of the primary fascicles; spermatangia rare, terminal to subterminal on primary or secondary fascicles, spherical, 6–9 µm in diameter; carpogonial branches straight, inserting on periaxial cells, short; involucral filaments nu-

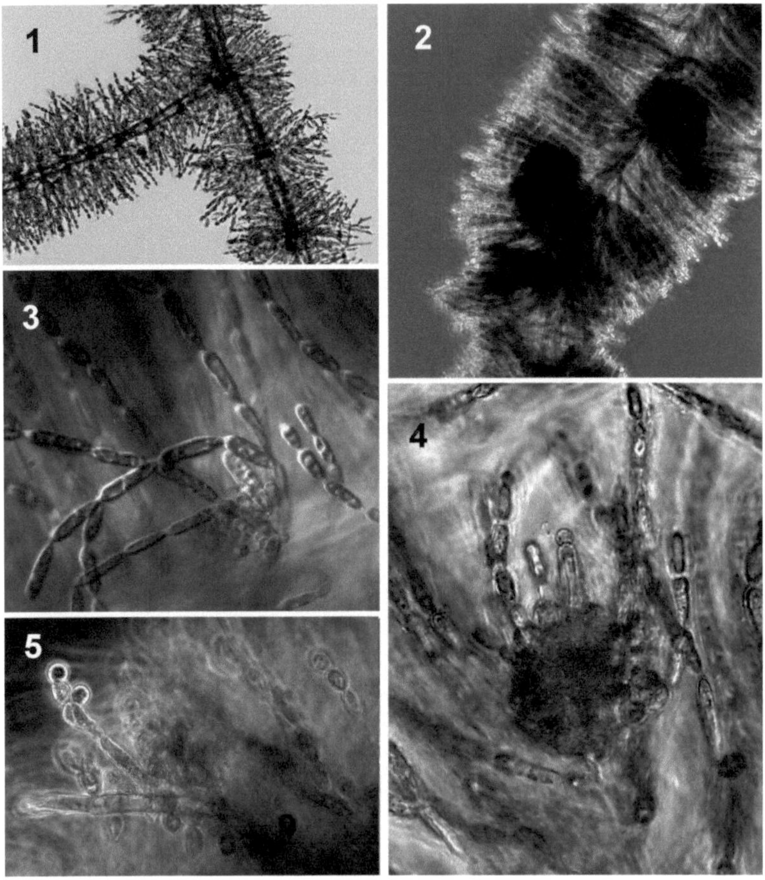

Plate 26: **Fig. 1–5.** *Batrachospermum keratophytum*: **1.** main axis with a branch (photo: J. Kostkevičienė); **2.** main axis with dense whorls and two large carposporophytes; **3.** young carpogonium-bearing branch between fascicles; **4.** developing carposporophyte after fertilization; **5.** prostrate gonimoblast filament with carposporangia.

Plate 27: **Fig. 1.** *Batrachospermum turfosum* (original authors, Sirodot 1884, ► Skuja 1969 for c and e): **a** – habitus of the plant; **b** – basal part of the thallus; **c** – primary and secondary fascicles; **d** – spermatia terminally on the fascicles; **e** – carpogonium-bearing branch; **f** – carposporangia **g–h** – monosporangia terminal on fascicles. **Fig. 2.** *Batrachospermum vogesiacum* (original author, Sirodot 1884): **a** – spermatangia terminal on fascicles; **b** –

1

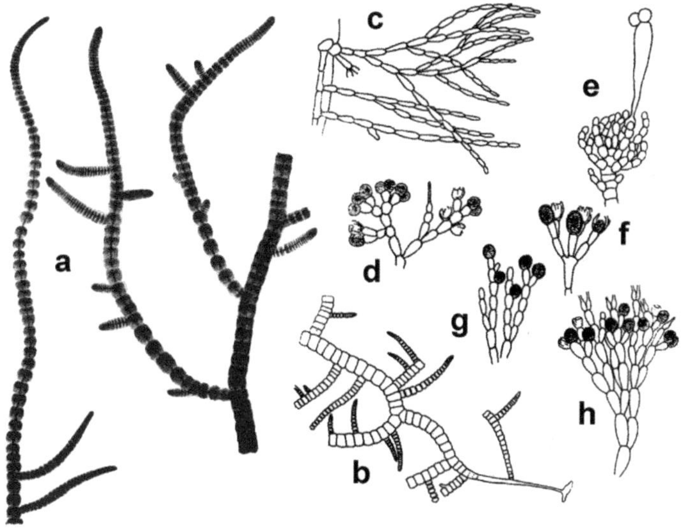

2

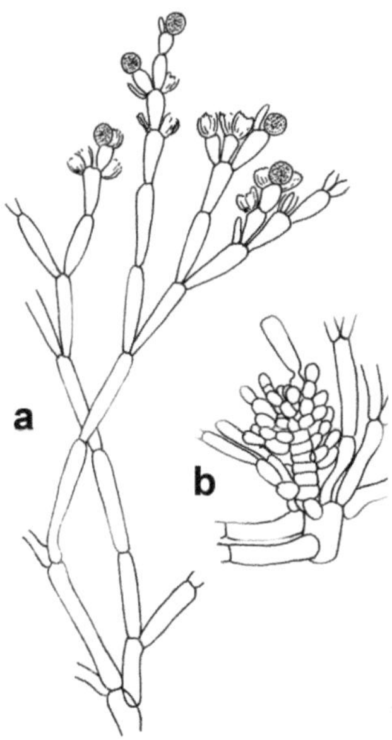

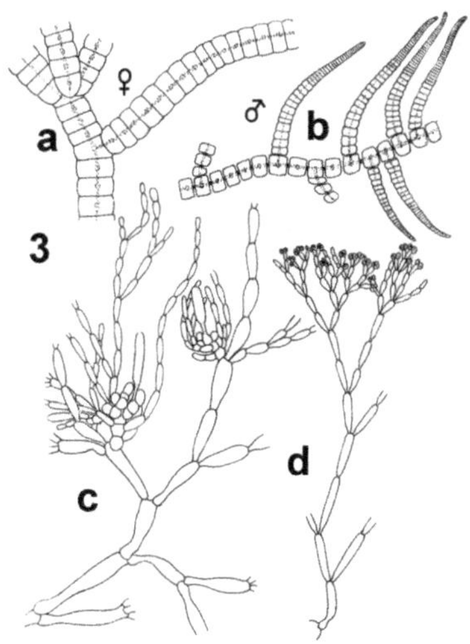

carpogonium-bearing branch arising from a periaxial cell. **Fig. 3.** *Batrachospermum elegans* (original author, Sirodot 1884): **a** – female plant with disc-shaped compressed whorls; **b** – male plant with less compressed whorls; **c** – fascicle with carpogonium-bearing branches arising from intercalary cells; **d** – terminal spermatangia on fascicles.

merous, short; carpogonium symmetrical, 35–60 µm long, with clavate trichogyne; trichogyne infrequently with basal protuberance; carposporophyte usually abortive, thus not observed; monosporangia numerous on the top of fascicles, 7.5–10 µm wide, 9–14 µm long.
Occurrence: In streams, springs, lakes (both clear-water and dystrophic lakes); commonly found in acidic waters, but also under neutral conditions all over Europe.

4. Section **Virescentia** Sirodot 1873

Section *Viridia* De Toni 1897, section *'Verts'* Sirodot 1884, secto *Claviformia* Reis 1973

Thallus intensely greenish, whorls well developed. Carpogonial branches short, straight, arising from pericentral or proximal fascicle cells. Carpogonia elongated with a cylindrical and pedicellate trichogyne. Carposporophytes along the main axis big, single or in pairs.

Important references: Israelson (1942); Starmach (1977); Necchi Jr. (1990); Necchi Jr. & Entwisle (1990); Sheath et al. (1994a); Kumano (2002).

Note: See note in section *Turfosa* concerning the position of *B. vogesiacum*.

Key to the species in section *Virescentia*:

1a	Carpogonium < 40 µm long	**2**
1b	Carpogonium > 40 µm long	**5**
2a	Carpogonium 20–40 µm long	**1. B. vogesiacum**
2b	Carpogonium 20–30 µm long	**3**
3a	Carposporophyte 150–300 µm wide, carpogonial branch long, up to 12 celled	**2. B. gulbenkianum**
3b	Carposporophyte < 180 µm wide	**4**
4a	Carpogonial branch short, composed of 3–5 cells, polyoecious	**3. B. azeredoi**
4b	Carpogonial branch long, up to 12-celled	**4. B. ferreri**
5a	Trichogyne clavate, having 1–3 basal knobs	**5. B. elegans**
5b	Trichogyne without basal knobs	**6**
6a	Carposporophytes numerous, carpogonial branch curved	**6. B. transtaganum**
6b	Carposporophytes axial, single, carpogonial branch not curved	**7. B. helminthosum**

1. **Batrachospermum vogesiacum** Schultz ex Skuja 1938 (plate **27**: fig. **2**)

Batrachospermum vagum var. *flagelliforme* Sirodot 1884, *B. flagelliforme* (Sirodot 1884) Necchi Jr. 1988, nom. illeg.

Plants monoecious (dioecious or monoecious according to Necchi Jr. 1990), very mucilaginous, 2–10 cm high, 340–700 µm wide, richly branched, grey, brownish olive or pale green. Whorls elliptical or bulgy, separated or coalescent, primary fascicles of 6–10 cell storeys; hairs short, frequent; secondary fascicles abundant, reaching the length of the primary fascicles. Spermatangia spherical, 5–6 µm in diameter, terminal on both primary and secondary fascicles. Carpogonial branches straight, arising from periaxial cells; involucral filaments numerous, short. Carpogonium 4–7 µm in diameter, trichogyne elongate-conical or cylindrical, indis-

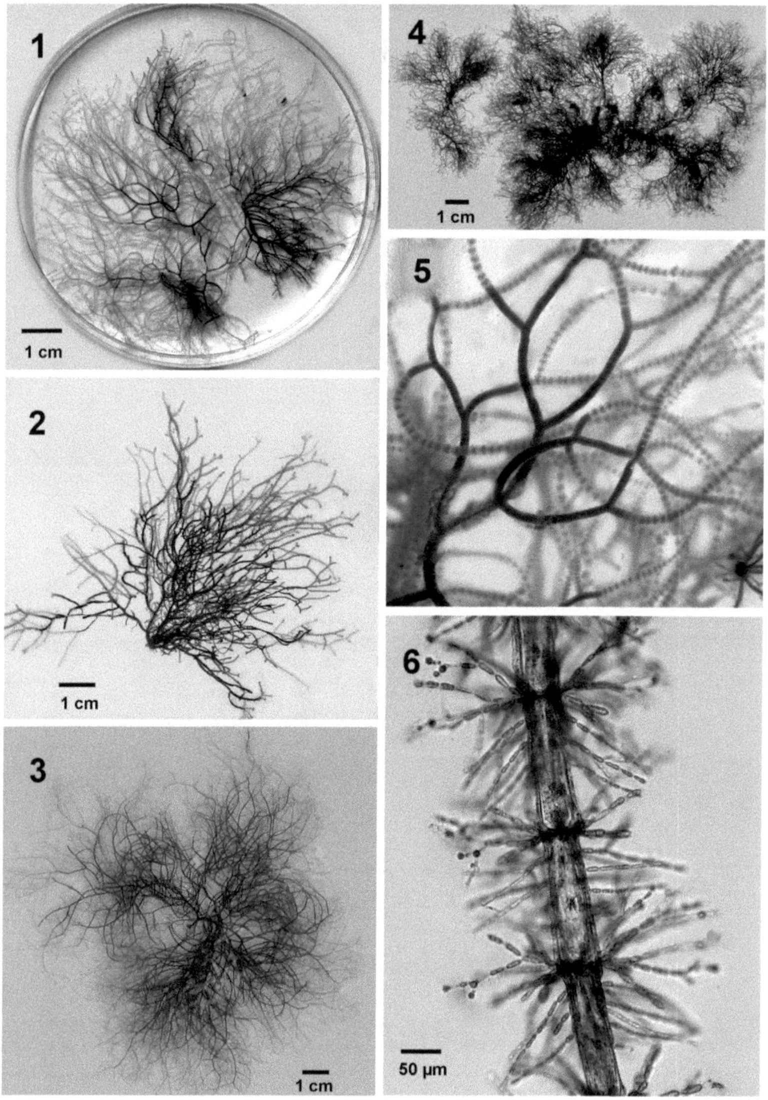

Plate 28: **Fig. 1–6.** *Batrachospermum turfosum*: **1–4.** habitus of the plant showing ramification; **5.** part of thallus with indistinct whorls and more distinct whorls in the younger branches; **6.** young part of the main axis showing primary and secondary fascicles, monospores terminal on the primary fascicles and axial cells with few cortical filaments.

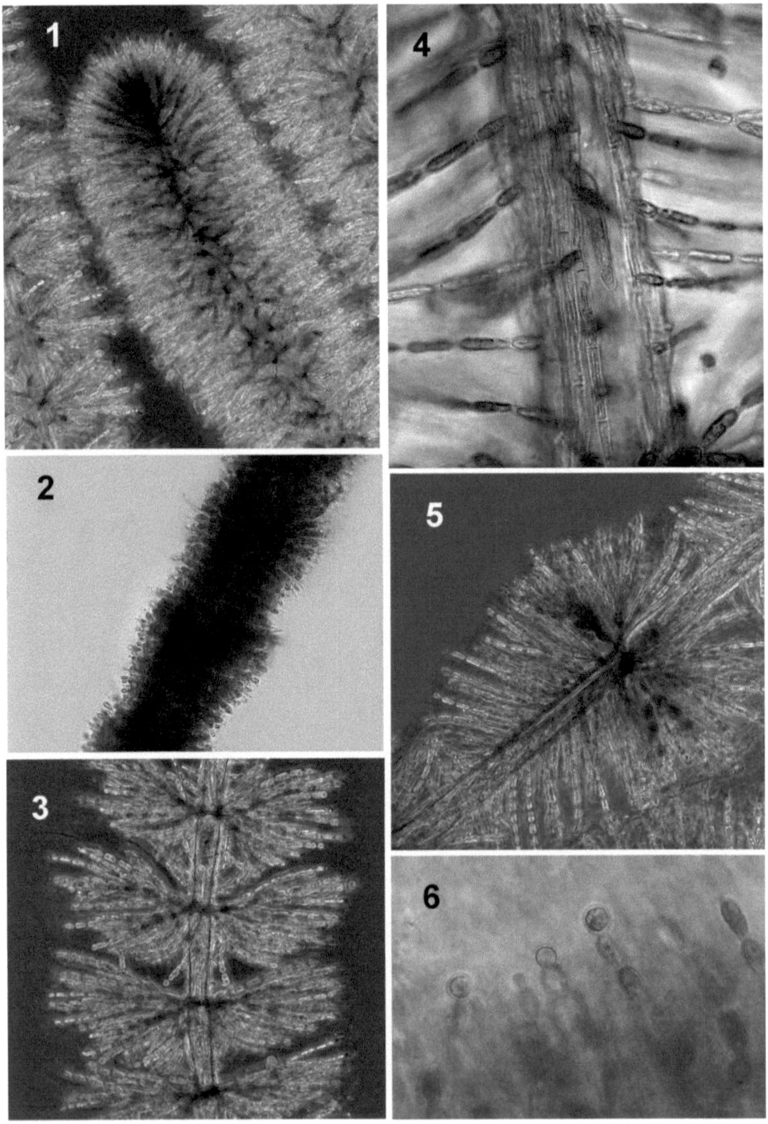

Plate 29: **Fig. 1–6.** *Batrachospermum turfosum*: **1.** apex of branch; **2.** part of thallus with confluent whorls; **3.** part of a young branch with plain main axis and separate whorls; **4.** main axis with cortical filaments and secondary fascicles; **5.** main axis with obconical whorl; **6.** monospores terminal on fascicles.

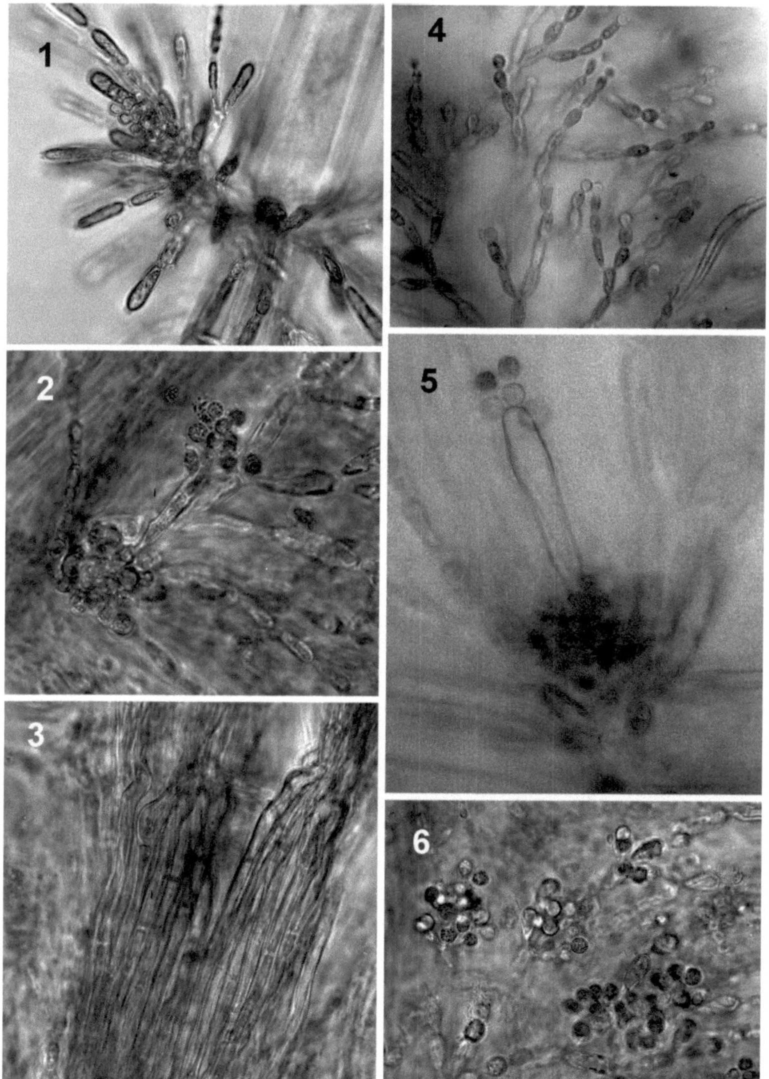

Plate 30: **Fig. 1–6.** *Batrachospermum turfosum*: **1.** main axis and apex of a new branch; **2, 5.** carpogonium-bearing branch with numerous short involucral filaments and long trichogyne attached by group of spermatia; **3.** cortical filaments with some bulbous cells; **4.** monosporangia at top of fascicles; **6.** groups of spermatangia at tops of fascicles.

tinct or shortly stalked. Carposporophytes single, sometimes in pairs, globose, 140–280(–330) µm wide, centrally inserted; carposporangia ellipsoidal, 8.5–13(–17) × 13–19(–25) µm. Monosporangia absent.
Occurrence: In peat-bog creeks and lakes (Israelson (1942); Europe: Belgium, France, Sweden.

2. Batrachospermum gulbenkianum Reis 1965 (no figures)

Plants monoecious, 4–5 cm high, isolated or in tufts, intense green in shaded habitats, yellow-green at sunny locations, mucilaginous. Branching irregular, branches usually at right angles, very slender. Primary fascicles scarce, only rarely numerous, simple, long and flagelliform. Secondary fascicles usually simple, flagelliform, less often branched. Whorls distinct, spherical to elongated ellipsoidal, slightly confluent, and radiating in all directions. Cortical filaments cylindrical and very coherent. Hairs short to long, with slightly inflated base, variable in abundance. Spermatangia at the top of fascicles and very abundant at involucral filaments. Carpogonial branches upright resembling small branch inserted on periaxial cells with dense involucral filaments enclosing carpogonium. Carpogonium truncate, 5–6.6 µm in diameter with very short-stalked trichogyne, truncate, rarely cylindrical, generally clavate, 20–33 µm long. Carposporophytes 1–3 more or less central, spherical or hemispherical, 150–300 µm in diameter. Carposporangia generally obovoidal, rarely spherical, 6.5–10 µm in diameter, 10–16.5 µm long.
Occurrence: Known only from Rio Alfusquiero, tributary of river Àgueda in Portugal.

Note: According to Reis (1965a) this species belonged to the section *Turfosa*. Kumano (2002) assigned it to section *Virescentia* due to radially branched and determinate gonimoblast filaments.

3. Batrachospermum azeredoi Reis 1967 (no figures)

Plants monoecious, 5–6 cm in height, blue-gray, moderately mucilaginous, enlarged at the base by cortical filaments. Whorls generally separated, terminal hairs are rare, short and slender, inflated at the base. Secondary fascicles absent or rare. Spermatangia occur at the top of fascicles, 4.0–6.6 µm long. Cortical filaments consist of large cylindrical cells. Carpogonial branches inserted on periaxial cells and basis of primary fascicles; carpogonium enclosed with short involucral filaments. Trichogyne sessile or with short stalk, clavate or cylindrical, 23–33 µm long. Carposporophytes usually rare, centrally inserted, 1–2 at each whorl, spherical to hemispherical, 120–180 µm in diameter. Carposporangia obovoidal, 8.3–10.0 µm in diameter, 13.0–16.6 µm long.
Occurrence: On river rocks from various localities in Portugal.

4. Batrachospermum ferreri Reis 1967 (no figures)

Plants polyoecious, 4–7 cm high, intensely green, very mucilaginous, solitary or in small tufts. Whorls truncate at their base, branches irregular and frequent, inserting at right angles in all directions. Secondary branches simple, flagelliform or sparsely branched. Male plants: whorls of contiguous and compressed elongate-ellipsoidal shape at the main axis, but separated and spherical or hemispherical on slender branches. Terminal hairs rare, short to long, slightly inflated at their base. Cortical filaments cylindrical, abundant in the older parts of the plants. Spermatangia occurring at proximal parts of fascicles. Female plants: whorls distinct, spheri-

cal to elongated-ellipsoidal. Terminal hairs are rare and very short. Secondary fascicles eiher missing or rare. Cortical cells like in male plants. Carpogonial branches short, usually arising from periaxial cells, rarely from primary fascicles; producing from the upper half to third part short involucral filaments enclosing the carpogonium and from the lower part unilateral filaments. Carpogonium truncate-cupula-shaped with a shortly stalked, clavate or truncate, rarely cylindrical 20–33 µm long trichogyne. Carposporophytes central, rare, 1–2 in each whorl, irregularly hemispherical, 130–180 µm in diameter. Carposporangia obovoidal, 6.6–10.0 µm in diameter, 10.0–13.3 µm long. Bisexual plants resembling male plants.
Occurrence: Known only from streams in Portugal.

5. Batrachospermum elegans Sirodot 1884 emend. Sheath, Vis et Cole 1994 (plate **27**: fig. **3**)

B. coerulescens Sirodot 1884, non *B. coerulescens* Bory 1823

Note: According to Necchi Jr. (1990), *B. coerulescens* Sirodot is synonymous to *B. helminthosum* Bory 1808.

Plants polyoecious, blue-green with brownish main axis, up to 9 cm high; whorls confluent, barrel-shaped, 316–1002 µm in diameter, secondary fascicles sparse. In male plants, whorls usually separated from each other; hairs sparse. Carpogonial branches consisting of 1–3 short cells, arising from either periaxial cells or proximal cells of the fascicle branch. Carpogonium 35.5–65.4 µm long with pedicellate, cylindrical, clavate trichogyne having 1–3 basal knobs or branches. Carposporophytes axial, 135–317 µm in diameter. Carposporangia obovoidal, 6.6–14.6 × 9.9–23.5 µm.
Occurrence: Known for Europe only from sandy bottom streams in France.

Note: Sheath et al. (1994a) assigned it as monoecious.

6. Batrachospermum transtaganum Reis 1970 (plate **31**: fig. **1**)

Plants polyoecious (according to Starmach (1977) monoecious), 2–3 cm high, light-green when growing in sunlight, dark-brown at shaded places, very mucilaginous. Branches inserted irregularly at almost right angles, showing in all directions. Male plant: whorls separated, spherical, but discoid towards the tips; secondary fascicles none or rare. Terminal hairs numerous, long and slender, inflated at the base. Spermatangia rare, produced both in the interior and at the top of fascicles. Female plant: whorls separate or contiguous, compressed (in winter), conical, disc-shaped towards the tips. Secondary fascicles numerous, mainly occupying the upper part of the internode, becoming shorter and shorter towards the lower whorl. Terminal hairs rare and short. Cortical filaments regular, cylindrical and yellowish. Carpogonial branches usually curved, arising from the periaxial cells, producing from the upper 2 to 3 cells short involucral filaments, which enclose the carpogonium and from the lower cells one or more longer filaments. Carpogonium with ovoidal or elongated-ellipsoidal trichogyne, pedicellate, cylindrical, 46–60 µm long. Carposporophytes numerous, green, irregularly hemispherical, 165–220 µm in diameter. Carposporangia obovoidal, 10.0–11.0 × 16.6–18.0 µm.
Occurrence: Known from two locations in the River Mira in Portugal.

Note: Entwisle et al. (2009) formed a new genus *Kumanoa* and transferred species from the sections *Hybrida* and *Contorta* to that new genus. Those species have curved or twisted carpogonial branches like in *B. transtaganum*, but only one (rarely two) large hemispherical carposporophytes. Thus *B. transtaganum* may have to be transferred to the new genus after further studies have been conducted.

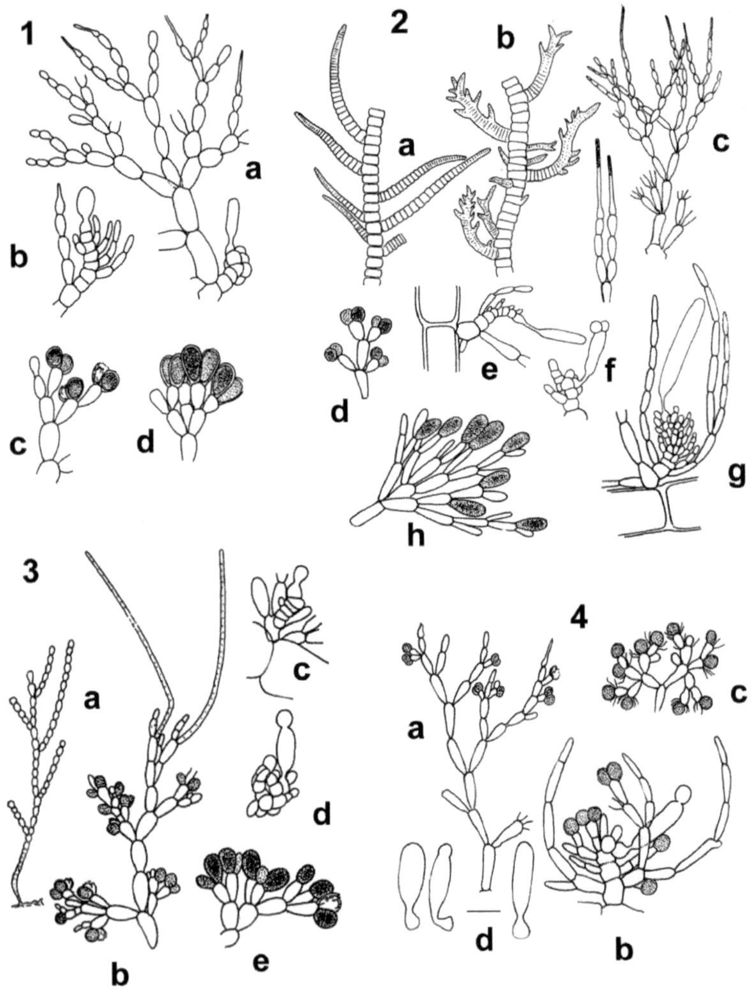

Plate 31: **Fig. 1.** *Batrachospermum transtaganum* (author Starmach (1977) according to photo by Reis 1970): **a** – primary fascicle and carpogonium-bearing branch arising from periaxial cell; **b** – curved carpogonium-bearing branch**;** **c** – spermatangia terminal on fascicle; **d** – carposporangia. **Fig. 2.** *Batrachospermum helminthosum* (original authors, Sirodot 1884 and Kylin 1912): **a** – habitus of a thallus in summer; **b** – habitus in winter; **c** – primary fascicle; **d** – spermatangia; **e–f** – carpogonium-bearing branches; **g** – young carposporophyte; **h** – carposporangia. **Fig. 3.** *Kumanoa virgatodecaisneana* (original author, Sirodot 1884): **a** – general view of thallus; **b** – fascicle with spermatangia and terminal hairs; **c–d** – carpogonium-bearing branches; **e** – carposporangia. **Fig. 4.** *Kumanoa globospora* (original author, Is-

7. Batrachospermum helminthosum Bory 1808 emend. Sheath, Vis et Cole 1994 (non *B. helminthosum* Sirodot 1884) (plate **31**: fig. **2**; plate **32**: fig. **1–5**; plate **33**: fig. **1–4**)

Batrachospermum graibussoniense Sirodot 1884, *B. bruziense* Sirodot 1884, *B. testale* Sirodot 1884, *B. virgatum* Sirodot 1884, *B. sirodotii* Skuja ex Flint 1950

Plants polyoecious, thallus 2–7 cm high, green with brownish main axis, fairly mucilaginous; whorls 300–800 μm wide, spherical, elliptical or peltate depending on the location in the thallus. Spermatangia abundant in male plants, in bisexual plants spermatangia are located only in certain regions of the thallus; purely female plants are rarely found. According to authors' descriptions, carpogonial branches are either composed of 1–5 short cells (Sheath et al. 1994) or of 3–9(–11) cells (Starmach 1977) and arise from a periaxial or proximal cell of the primary fascicle; carpogonium (25–)40–79 μm long with a pedicellate or cylindrical trichogyne. Carposporophytes axial, single, 120–415 μm in diameter, 102–420 μm high, densely clustered. Carposporangia obovoidal, 5.3–16.9 × 9.8–27.6 μm.
Occurrence: In lowland rivers; occurs all over Europe.

Note: Sheath et al. (1994a) assigned it as monoecious.

2. **Kumanoa** Entwisle, Vis, Chiasson, Necchi et Sherwood 2009

Batrachospermum section *'Hybride'* Sirodot 1884, section Hybrida De Toni 1897, section *Contorta* Skuja 1931 emend Vis et Entwisle 2000

Similar to *Batrachospermum* (L.) DC. emend Entwisle, Vis, Chiasson, Necchi et Sherwood (2009), but carposporophytes are always found in the center of the whorl and have usually a hemispherical shape, gonimoblast filaments always determinate; carpogonial branches curved, twisted or spirally coiled, with cells clearly differentiated from vegetative cells.

Important references: Hamel (1925a); Skuja (1928); Israelson (1942); Starmach (1977); Necchi Jr. (1990); Sheath et al. (1992); Sheath & Vis (1995); Kumano (2002); Entwisle et al. (2009)

Note: Entwisle et al. (2009) transferred 17 species from *Batrachospermum* to this genus, 2 of which are found also in Europe. In addition to those, Reis (1965b, 1972b, 1973) described four species of the sections *Hybrida* and *Contorta*, which have been found in the type localities only (Portugal). Those should be regarded as *Kumanoa* species, but due to lack of genetic studies their names are unchanged in this key. These species are described after the known *Kumanoa* species.

Key to the species:

1a Carpogonial branches curved, carposporophytes 100–120 μm in diameter . . . . **1. K. virgatodecaisneana**

1b Carpogonial branches strongly undulated, carposporophytes 160–350 μm in diameter . . . . **2. K. globospora**

---

raelson 1942): **a** – spermatangia terminally located on fascicles; **b** – curved carpogonium-bearing branch with spermatangia on involucral filaments; **c** – loose carposporophyte with carposporangia terminally inserting on gonimoblast filaments; **d** – different shapes of trichogynes.

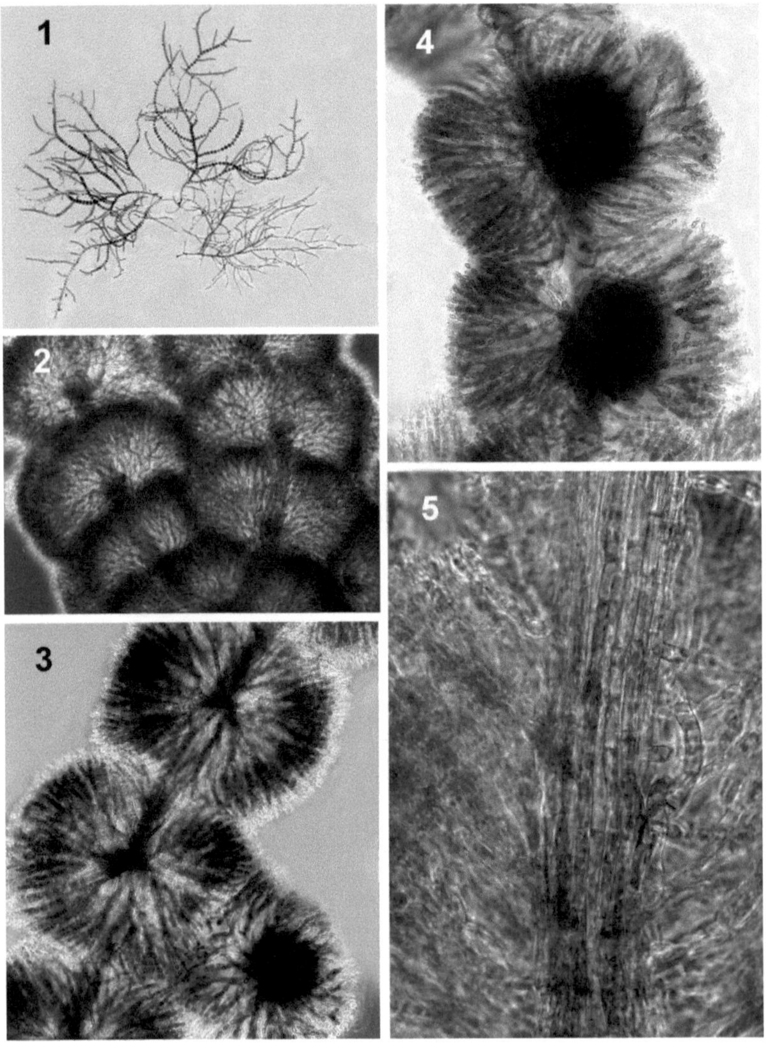

Plate 32: **Fig. 1–5.** *Batrachospermum helminthosum*: **1.** habitus showing ramification; **2–3.** compressed and spherical whorls; **4.** whorls with hemispherical, large carposporophytes; **5** – regular cortication on the main axis.

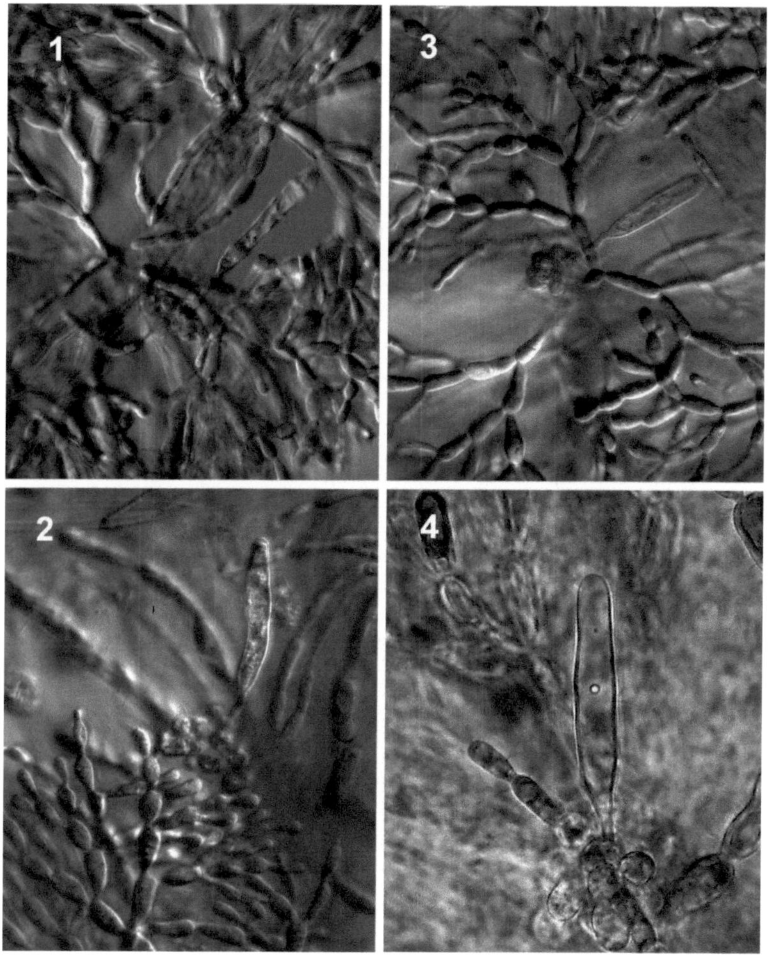

Plate 33: **Fig. 1–4.** *Batrachospermum helminthosum*: carpogonium-bearing branches and carpogonia with long stalked trichogynes.

1. Kumanoa virgatodecaisneana (Sirodot) Entwisle, Vis, Chiasson, Necchi et Sherwood 2009 (plate **31**: fig. **3**)

Batrachospermum virgato-decaisneanum Sirodot 1884, *B. mikrogyne* Skuja ex Flint 1953

Thallus 2–8 cm high, very mucilaginous, intensely green; whorls spherical or elliptical, separated or coalescent, 300–390 µm wide (American populations 274–686 µm), with short cylindrical, obovoidal or elliptical cells; terminal hairs sparse to frequent. Secondary fascicles more or less frequent, finally covering all internodes, reaching the length of the primary fascicles. Spermatangia spherical to ovoid, ca. 6 µm in diameter, located terminally on fascicles at different levels. Carpogonial branches arising from periaxial cells. Carpogonium 15–25 µm long, trichogyne short, ellipsoidal, cylindrical, lanceolate, indistinctly stalked. Carposporophytes single (rarely in pairs), spherical or hemispherical, 100–120 µm in diameter, placed in the center of the whorl. Carposporangia ellipsoidal, obovoidal or pear-shaped, 8–10 × 10.5–14.0 µm. Monosporangia not known.
Occurrence: In running and standing waters, clay-pit and peat bogs; Europe: France, Latvia, Sweden.

2. Kumanoa globospora (Israelson) Entwisle, Vis, Chiasson, Necchi et Sherwood 2009 (plate **31**: fig. **4**; plate **34**: **1–6**)

Batrachospermum cipoense Kumano et Necchi 1985, *B. jolyi* Necchi 1986, *Batrachospermum globosporum* Israelson 1942

Plants monoecious, 2–7 cm high, very mucilaginous, green with bluish shade. Branchings sparse to rich, axis generally distinct. Branches cylindrical, apices obtuse. Cortical filaments usually sparse. Whorls separated, ellipsoid, later after development of secondary fascicles confluent, 300–600 µm wide. Terminal hairs abundant and often long. Spermatangia abundant, located on tips of fascicles or scattered inside the whorl and also on the top of involucral filaments, spherical or obovoidal, 5.5–7.5 µm in diameter. Carpogonial branches arising from the periaxial cells, short, strongly spirally twisted. Involucral filaments numerous, short. Carpogonium asymmetrical, 20–30–(40) µm long, with shortly or indistinctly stalked, conical, sometimes obovoidal or cylindrical trichogyne. Carposporophytes sessile, single or often in pairs, centrally inserted, hemispherical, 160–350 µm in diameter, 110–160 µm high, loose. Carposporangia almost spherical, rarely obovoid, 7.5–11.0 µm in diameter, 8.5–13.0–(15.0) µm long. Monosporangia unknown.
Occurrence: In lakes; in Europe only at four localities in Sweden.

Key to Portuguese (Reis) species (as *Batrachospermum*):

1a Monosporangia present 2
1b Monosporangia absent 3
2a Monoecious; spermatangia rare, large and spherical, 10 µm in diameter, at the tips of primary and secondary fascicles . . . . **1. B. pseudocarpum**
2b Dioecious; spermatangia spherical, 6.6–8.0 µm in diameter, located along secondary fascicles . . . . **2. B. lusitanicum**
3a Carpogonial branches curved, carposporophytes rare, semi-spherical, 1 per whorl . . . . **3. B. henriquesianum**
3b Carpogonial branches straight to curved, carposporophytes numerous, spherical . . . . **4. B. abilii**

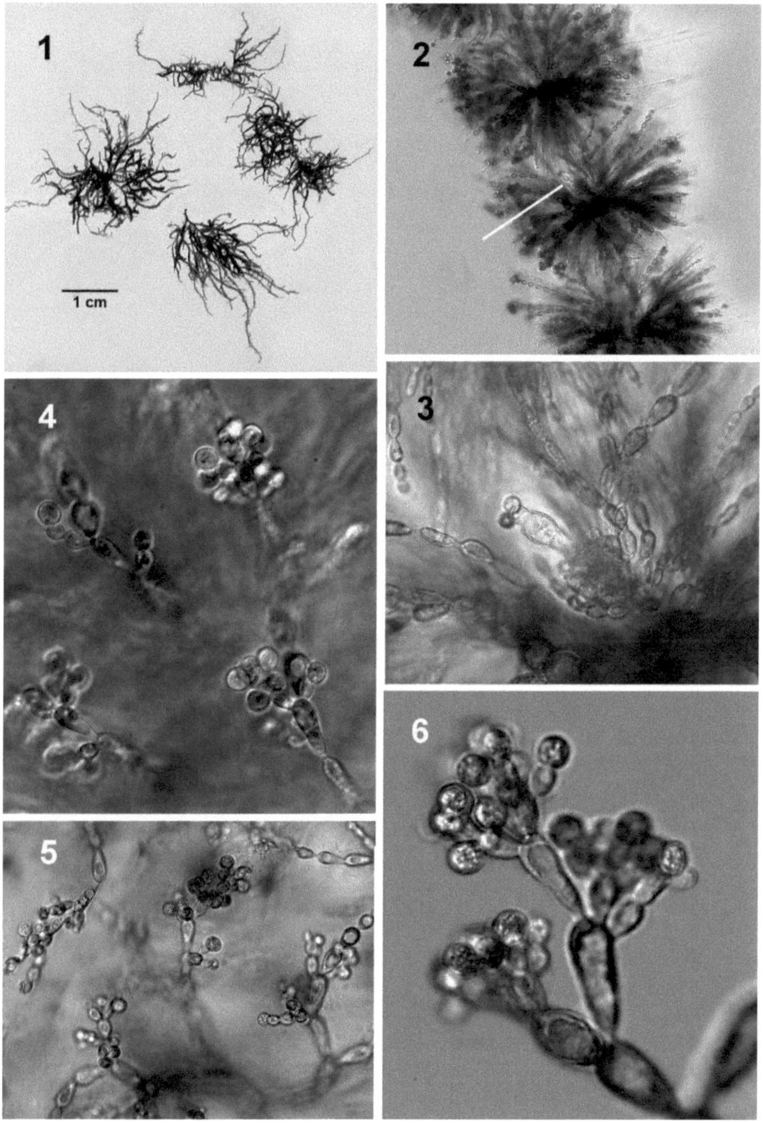

Plate 34: **Fig. 1–6.** *Kumanoa globospora*: **1.** habitus showing ramification; **2.** whorls with a carpogonium-bearing branch arising from periaxial cells; **3.** carpogonium-bearing branch and carpogonium with spermatia attached on top of trichogyne; **4–6.** groups of spermatangia terminally and subterminally located on fascicles.

1. Batrachospermum pseudocarpum Reis 1973 (plate **35**: 1–7; plate **36**: **1–6**)

Plants monoecious, 4–5 cm high, moss-green, very mucilaginous. Main axis densely branched, with branches at right angles. Whorls contiguous and obtruncate, secondary fascicles numerous immediately below the whorls, cortical filaments cylindrical. Terminal hairs rare, very short to long, inflated at their base. Spermatangia sporadic, large and spherical, 10 μm in diameter, located on top of primary and secondary fascicles. Carpogonial branches spirally twisted, inserted on periaxial cells. Shorter involucral filaments incurved thus appearing sphere shaped, longer ones are slightly curved and overtopping the former. Carpogonium asymmetrically truncate with shortly stalked, cylindrical trichogyne. True carposporophytes not observed, pseudo-carposporophytes fixed on the axis of a whorl, and also at the periphery of the whorl, semispherical or spherical, 200–300 μm in diameter. Monosporangium-bearing fascicles numerous, monosporangia obovoidal, 6–10 μm in diameter, 13–20 μm long, distributed on whorls.
Occurrence: Known from type locality, Ribeira da Louçainha, Espinhal Mountains in Portugal and from Central Finland, River Konttijoki, Saarijärvi.

Note: According to Kumano (2002), this species belongs to section *Contorta*, subsection *Intorta*. The newest unpublished results of genetic analyses indicated that this species could be conspecific with *Sirodotia suecica*.

2. Batrachospermum lusitanicum Reis 1965b

Plants dioecious, 3–4 cm high, at first pyramidal, afterwards confluent, forming tufts, moss-green, very mucilaginous. Branching corymbiform, rarely pyramidal. Main axis of male plants less branched than that of female plants (female plants are extremely branched). Whorls generally distinct, barrel-shaped. Terminal hairs short, inflated at their base. Secondary fascicles long in the upper part of the internode and becoming shorter when reaching the lower contiguous whorl. Spermatangia spherical, 6.6–8.3 μm in diameter, located at the base of whorls along secondary fascicles. Whorls generally contiguous and elongated discoidal, rarely indistinct at their base. Apices with terminal short hairs, which are inflated at their base. Cortical filaments abundant at lower thallus parts, cylindrical. Carpogonial branches spirally twisted, frequently inserted on periaxial cells, rarely on secondary fascicles and cortical filaments. Shorter involucral filaments incurved with a spherical shape, longer ones are slightly curved and overtopping the former. Carpogonium asymmetrically elongated-ovoidal or cylindrical with short stalked or sessile trichogyne. Carposporophytes 1 to 2 at each whorl, apparently fixed along the main axis, spherical, 150–300 μm in diameter. Carpospores obovoidal, 10–20 μm in diameter. Monosporangia formed on whorls at the plant base, obovoidal, 13–16.0 × 20.0–26.6 μm, distributed in whorls.
Occurrence: Known only from type locality, river Alfusqueiro, tributary of river Àgueda in Portugal.

Note: According to Kumano (2002), this species belongs to section *Contorta*, subsection *Intorta*.

3. Batrachospermum henriquesianum Reis 1972b

Plants monoecious, 1.5–2.5 cm high, light-green at sunny locations, violet in shaeded ones; very mucilaginous, solitary, perennial. Branching irregular, main axis attenuated towards the base, naked in the lower part. Apices of branches obtuse. Whorls indistinct, inclined forwards along spine-shaped fascicles. Secondary fas-

Plate 35: **Fig. 1–7.** *Batrachospermum pseudocarpum*: **1.** habitus of the plants; **2.** lower part of the plant with thalli radially growing from basal plate; **3.** filaments of the basal plate; **4.** obconical whorls from the middle part of thallus with pseudo-carposporophytes; **5.** regular cortical filaments and secondary fascicles; **6.** new branch with monosporangia terminally located on a fascicle; **7.** apex of thallus branch.

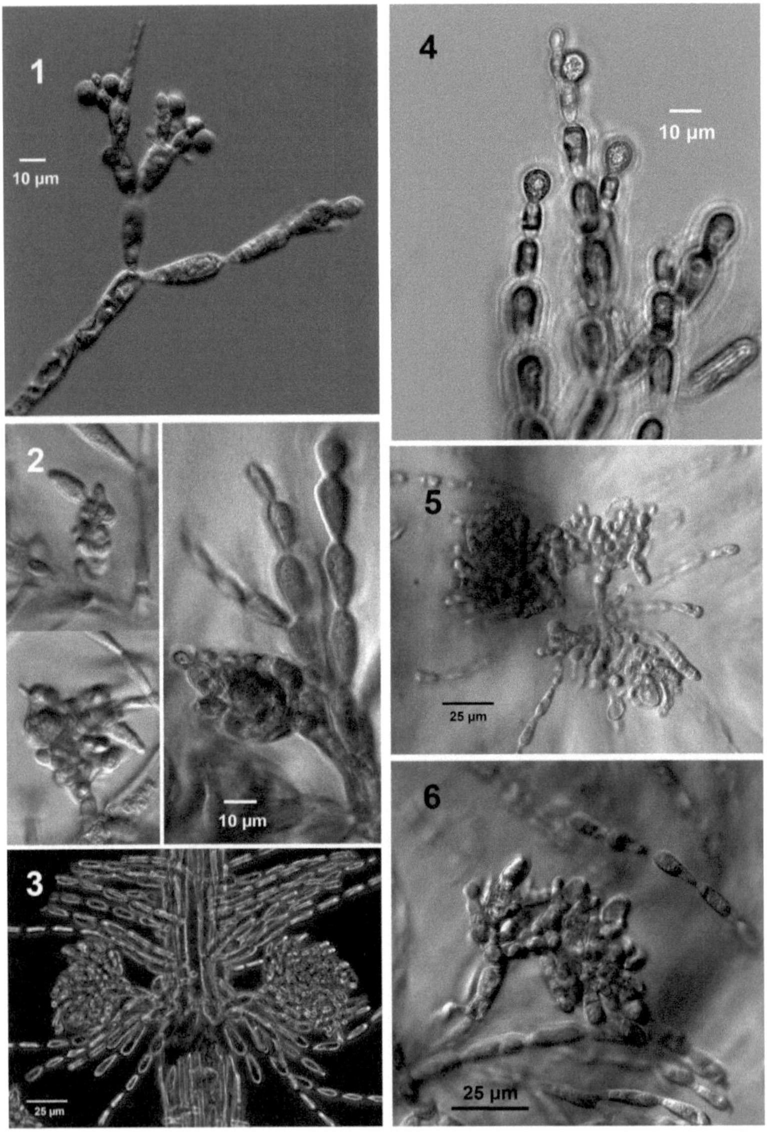

Plate 36: **Fig. 1–6.** *Batrachospermum pseudocarpum*: **1.** groups of spermatangia terminally located on a fascicle; **2.** young carpogonium-bearing branches; **3.** main axis with two pseudo-carposporophytes; **4.** monosporangia terminally located on fascicles; **5–6.** pseudo-carposporophytes.

cicles equal the radius of whorl. Cortical filaments very numerous, cylindrical, forming successive layers. Rarely, apices show short terminal hairs, inflated at their base. Spermatangia located at the periphery of primary and secondary fascicles, spherical, 6.6–8.6 µm in diameter. Carpogonial branches inserted on periaxial cells, curved. Involucral filaments short, incurved into a form of a sphere. Carpogonium asymmetric with short-stalked trichogyne. Carposporophytes quite rare, 1 per whorl, hemispherical, fixed on the main axis, 120–300 µm in diameter. Carpospores obovoidal or ellipsoidal, 26.7–30.0 µm in diameter, 40–46 µm long. Occurrence: Known only from type locality, spring Fonte velha in the village Eirol near Aveiro, Portugal.

Note: According to Kumano (2002), this species belongs to section *Contorta*, subsection *Torrida* Kumano.

4. **Batrachospermum abilii** Reis 1965b

Plants monoecious, 2–3 cm high, when young pyramidal, later making dense tufts, extremely mucilaginous, moss-green. Branching abundant, dense, irregular. Main axis disappearing among the upper branches, towards the base thickened by cylindrical and rather coherent cortical filaments. Whorls contiguous, discoid, first distinct but becoming indistinct after reproduction. Secondary fascicles numerous, gradually covering internodes. Spermatangia are located on apices of fascicles and on involucral filaments, as mature obovoidal 2.5–3.5 µm in diameter, 4–6 µm long. Carpogonial branches straight or curved, consisting of 2–8 cells, usually arising from primary fascicles, rarely from secondary ones. Involucral filaments short, enveloping; carpogonium cylindrical, 4.5–6.0 µm long, trichogyne obovoidal. Carposporophytes numerous, spherical, 150–300 µm in diameter, more than 1/3 of the radius of whorls. Carpospores obovoidal, 8–10 µm in diameter, 13–16 µm long. Occurrence: In Europe only at the type locality Vermoim, Aveiro, Portugal. Found also in Brazil under the name *B. virgato-decaisneanum* (O. Necchi Jr., personal comm.).

Note: According to Kumano (2002), the species belongs to section *Hybrida* De Toni.

3. **Sirodotia** Kylin 1912

Plants irregularly branched; carpogonium-bearing branches short, arising from periaxial cells or cells of fascicles or both. Carpogonia asymmetrical with elongated conical or club-shaped trichogynes. Carposporophytes indeterminate in shape, creeping along cortical filaments. Gonimoblast filaments diffuse and indeterminate, developing directly from the fertilized carpogonia.

Important references: Kylin (1912); Israelson (1942); Starmach (1977); Necchi Jr. & Entwisle (1990); Necchi Jr. et al. (1993); Vis & Sheath (1999); Kumano (2002); Xie & Shi (2004); Eloranta & Kwandrans (2002, 2007).

Key to the species:

1a Plants dark green or bluish-green, whorls ellipsoidal, obovoidal, finally pear-shaped or coalescent, monoecious . . . . **1. S. suecica**

1b Plants pale green or yellowish green, whorls narrow, obconical, internodes long, secondary fascicles short, gradually lengthened, dioecious . . . . **2. S. tenuissima**

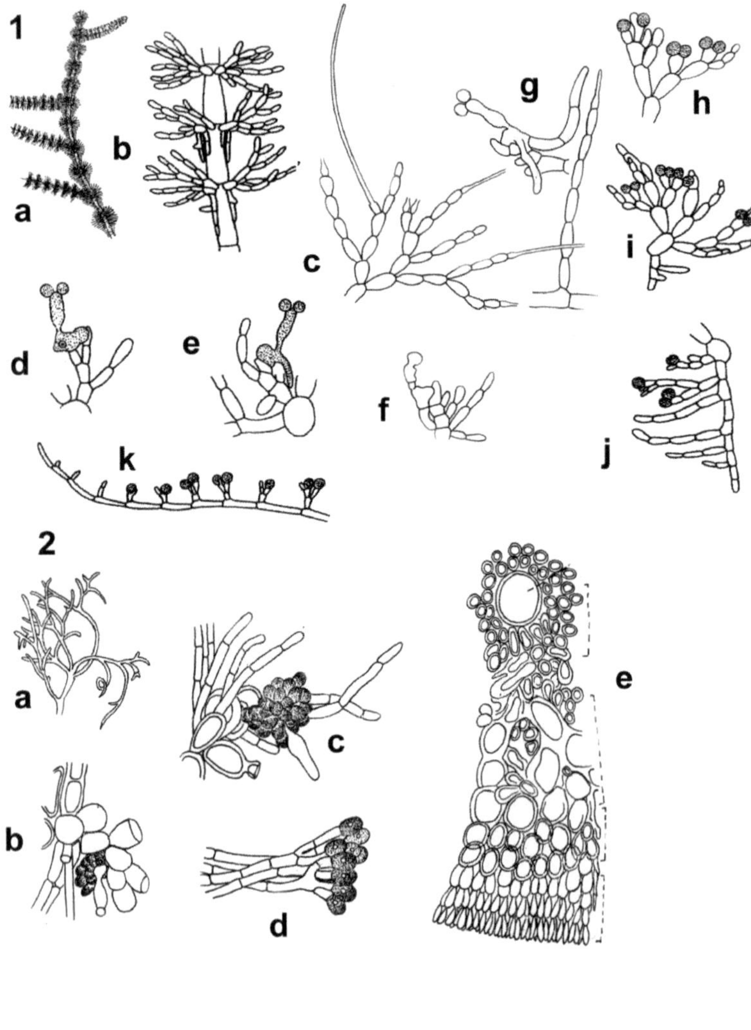

Plate 37: **Fig. 1.** *Sirodotia suecica* (original authors; Kylin 1912, Skuja 1931): **a** – main axis with truncate-pyramidal whorls; **b** – young main axis with whorls; **c** – primary fascicle with terminal hairs; **d–g** – fertilized carpogonium with attached spermatia; **h–i** – spermatangia terminal on fascicle; **j** – spermatangia terminal on secondary fascicles; **k** – a carposporophyte with an indeterminate prostrate gonimoblast filament and carposporangia. **Fig. 2.** *Tuomeya americana* (original author, Setchell 1890): **a** – habit of the thallus; **b–c** – young carposporophytes; **d** – carposporangia terminal on gonimoblast filaments; **e** – cross section of thallus showing axial cell, cortical filaments and fascicles.

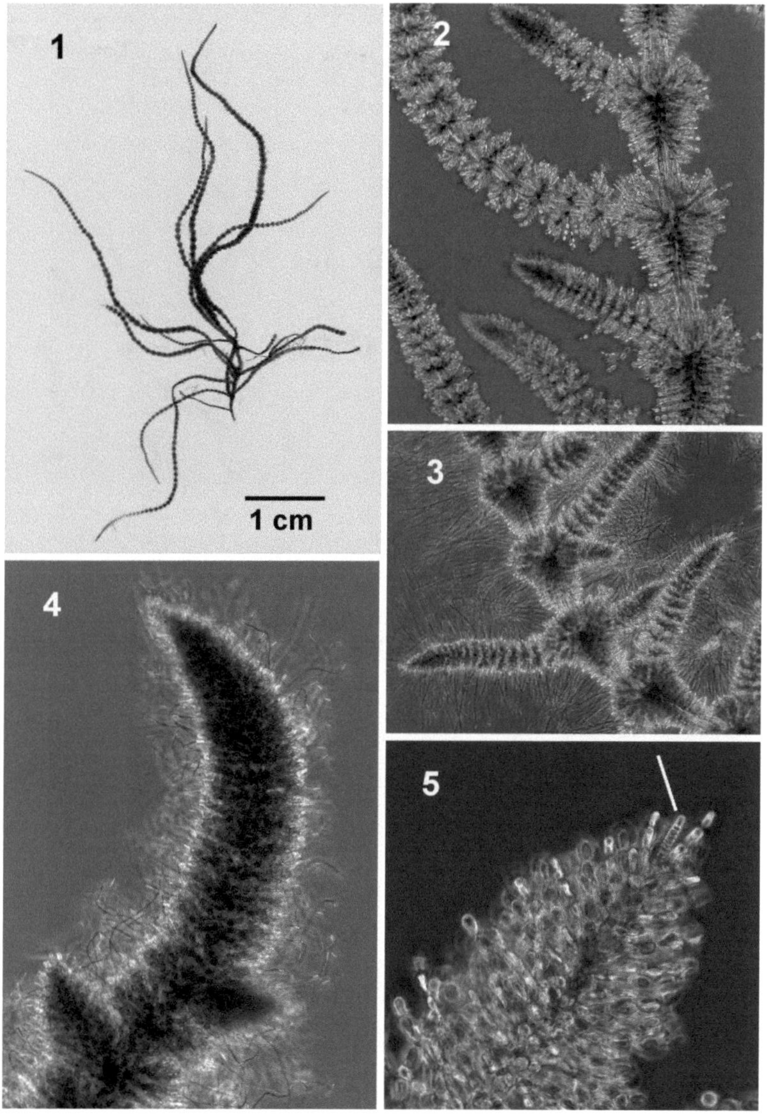

Plate 38: **Fig. 1–5.** *Sirodotia suecica*: **1.** habitus of a sparsely branched plant; **2–3.** main axis with obconical whorls, young branches with immature whorls and abundant hairs located terminally on fascicles (3); **4.** young branch; **5.** typical apex of the new branch.

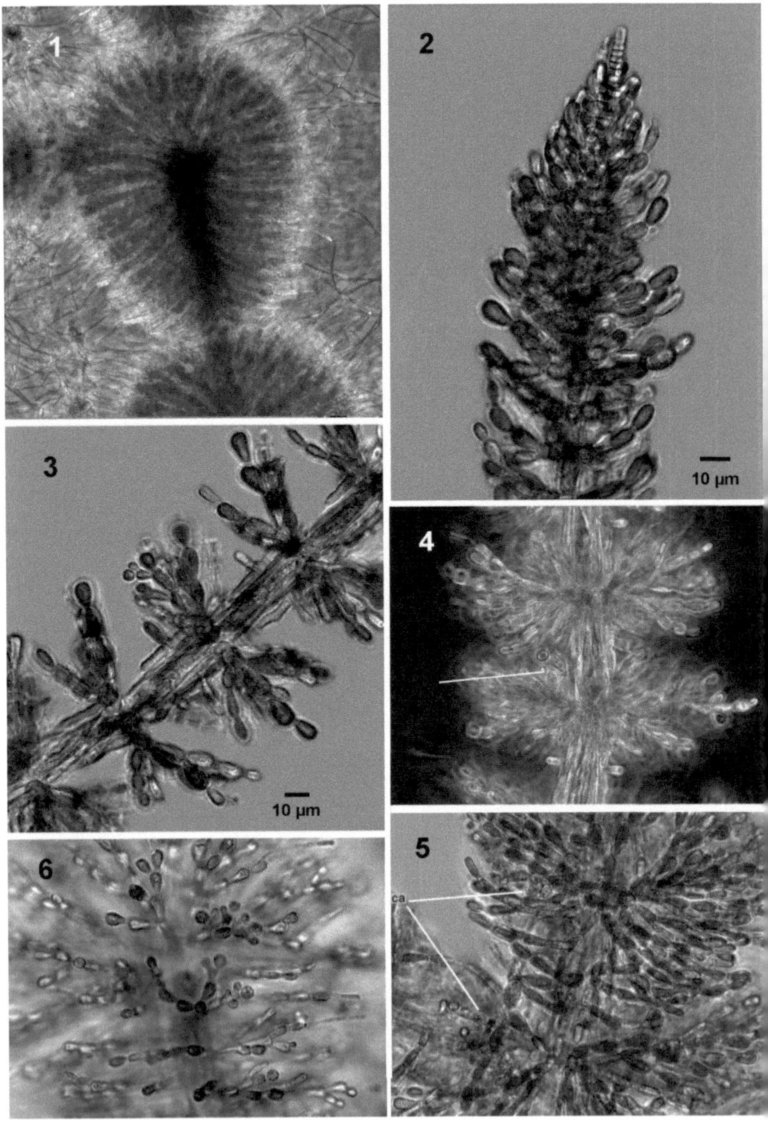

Plate 39: **Fig. 1–6.** *Sirodotia suecica*: **1.** main axis with obconical whorls and abundant hairs inserting terminally on fascicles; **2.** apex of young branch; **3.** young branch with immature whorls and spermatangia located terminally on fascicle; **4–5.** fascicles with carpogonia (ca) and single spermatangium; **6.** spermatangia situated terminally on fascicles.

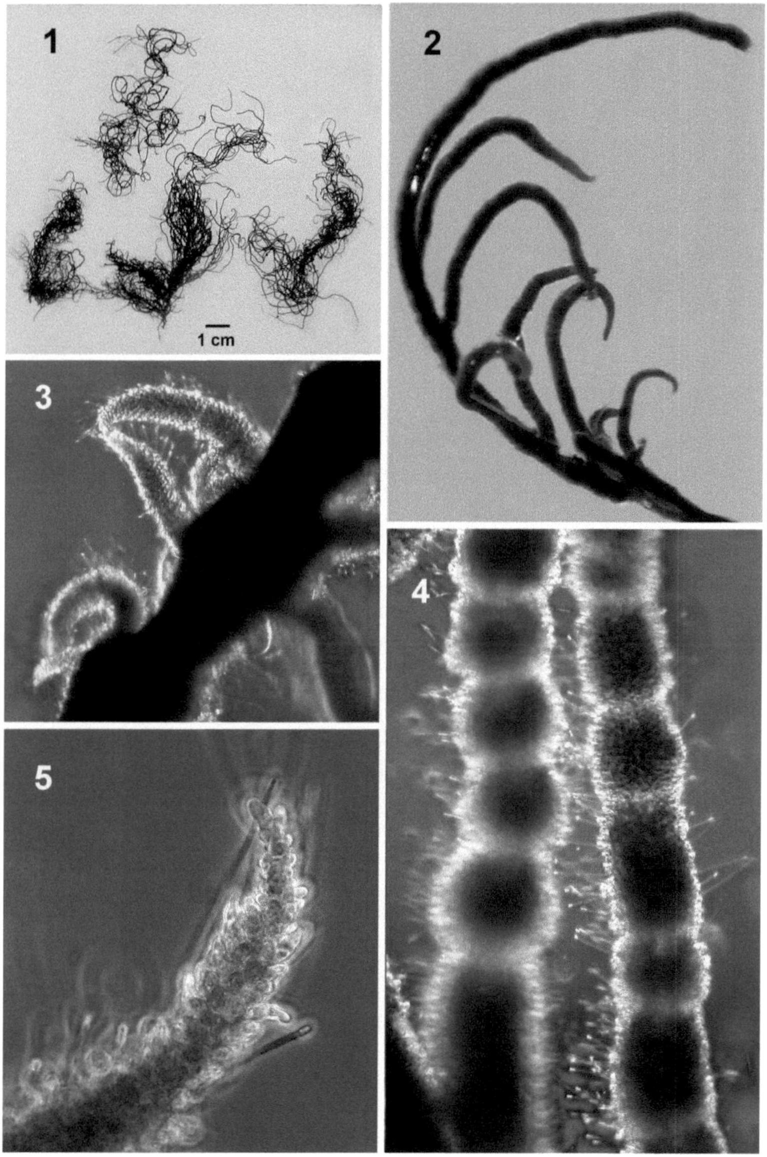

Plate 40: **Fig. 1–5.** *Sirodotia suecica*: **1–2.** habitus of plants growing in fast current with invisible whorls; **3.** main axis with very dense whorls and young branches; **4.** axes with dense confluent whorls; **5.** apex of young branch and some terminal hairs of fascicles.

Plate 41: **Fig. 1–5.** *Sirodotia tenuissima*: **1.** main axis and branches of the thallus; **2.** one internode of the main axis with numerous short secondary fascicles and young branches starting from node; **3.** main axis of a young branch with short secondary fascicles and two carpogonia

Note: According to Vis and Sheath (1999), these taxa are conspecific from a genetic perspective and should be unified. In Finnish material, those two taxa were occurring together on the same stone, but could easily be distinguished (Eloranta & Kwandrans 2002, 2007). We therefore kept the species in the actual key.

1. Sirodotia suecica Kylin 1912 (plate **37**: fig. **1**; plate **38**: fig. **1–5**; plate **39**: fig. **1–6**; plate **40**: fig. **1–5**)

Sirodotia fennica Skuja 1931, *Sirodotia acuminata* Skuja in Flint 1950, *Batrachospermum vagum* (Roth) C. Agardh f. *tenuissimum* Collins 1895

Plants monoecious, 2–12 cm high, dark green or bluish green. Whorls typically elliptical, 200–300 µm wide, separated from each other in young branches, later coalescent. Terminal hairs present. Secondary fascicles numerous, finally as long as primary ones. Spermatangia spherical, 5–6 µm in diameter, terminally located on fascicles. Carpogonium-bearing branches short, consisting of 2–5 barrel-shaped cells arising from periaxial cells. Carpogonia with a hemispherical protuberance on one side of the basal part, 30–40 µm in length, trichogyne ellipsoidal, finally cylindrical or spatula-shaped and more or less distinctly stalked. Carposporophytes indefinite in shape, gonimoblast filaments irregularly branched, initiating from the basal protuberance of carpogonium, creeping along cortical filaments. Carposporangia obovoidal to elongated pear-shaped, 6–8 µm wide and 8–12 µm long.
In fast flowing environments, the fascicles may develop very densely making whorls completely confluent. However, even then branch apices are gradually tapering to sharp ends, where the distinct tip of apical cells can be seen.
Occurrence: In running waters on stones, wood or mosses, sometimes also attached on older parts of *Batrachospermum*. Europe: Rather common in northern Europe (Finland, Sweden and Russia). Known also from Scotland (Loch Cluanie outflow).

2. Sirodotia tenuissima (Collins) Skuja ex Flint 1948 (plate **41**: fig. **1–5**)

Plants dioecious; up to 15 cm long, delicate, yellowish-green filaments with long internodes; secondary fascicles gradually elongated along internode, terminal hairs absent. Structure of carpogonium, carposporophyte and gonimoblast filaments like in *Sirodotia suecica*. Carpospores ovoidal.
Occurrence: Known from one location in Central Sweden and from three locations in Finland.

Note: The finding of *Sirodotia huillensis* as new species for Europe (Del Grosso 1981) has to be scrutinized; it probably belongs to *Batrachospermum atrum* (Hudson) Harvey.

4. Tuomeya Harvey 1858

Batrachospermum section *Tuomeya* (Harvey) Necchi et Entwisle 1990

Plants irregularly branched, cartilaginous, continuous and solid. Plants composed of a longitudinal uniaxial filament, cortical filaments around it form a pseudoparenchymatic cortex and dense whorls. Monospecific genus.

(white lines); note the characteristic hemispherical protuberance in the basal part of carpogonium; **4.** apex of young branch; **5.** main axis with abundant secondary fascicles.

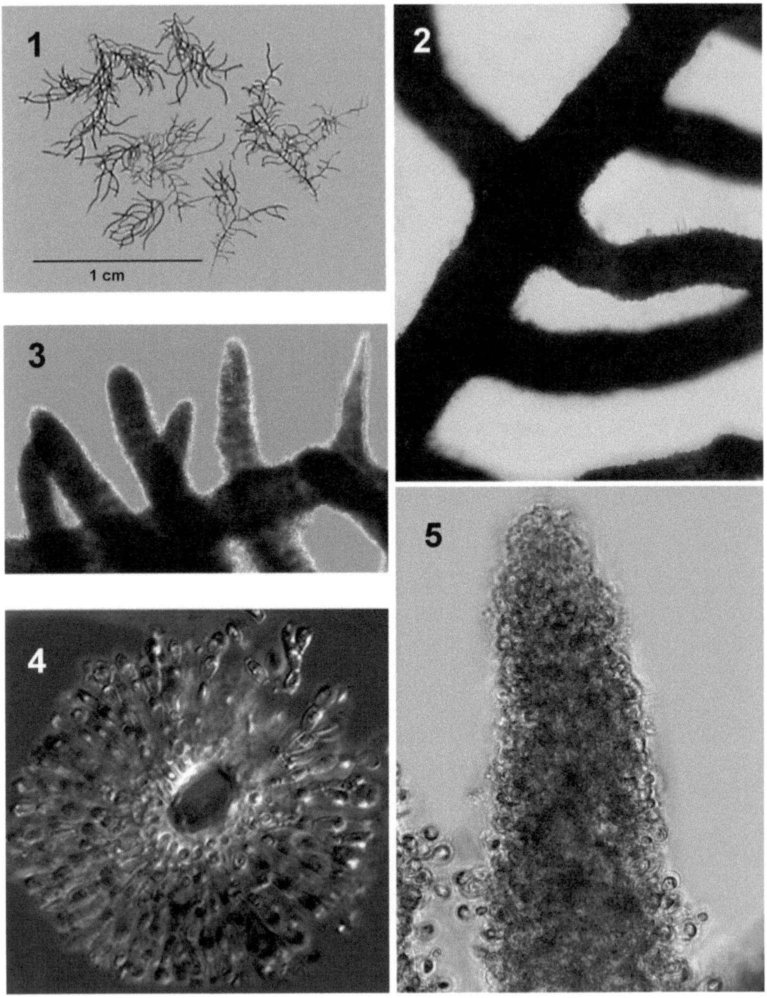

Plate 42: **Fig. 1–5.** *Tuomeya americana*: **1.** – habitus of plants; **2–3.** – main axis with dense confluent whorls and typical ramification; **4.** – cross-section of thallus showing axial cell, cortical layer around it and dense fascicles; **5.** – apex of a branch.

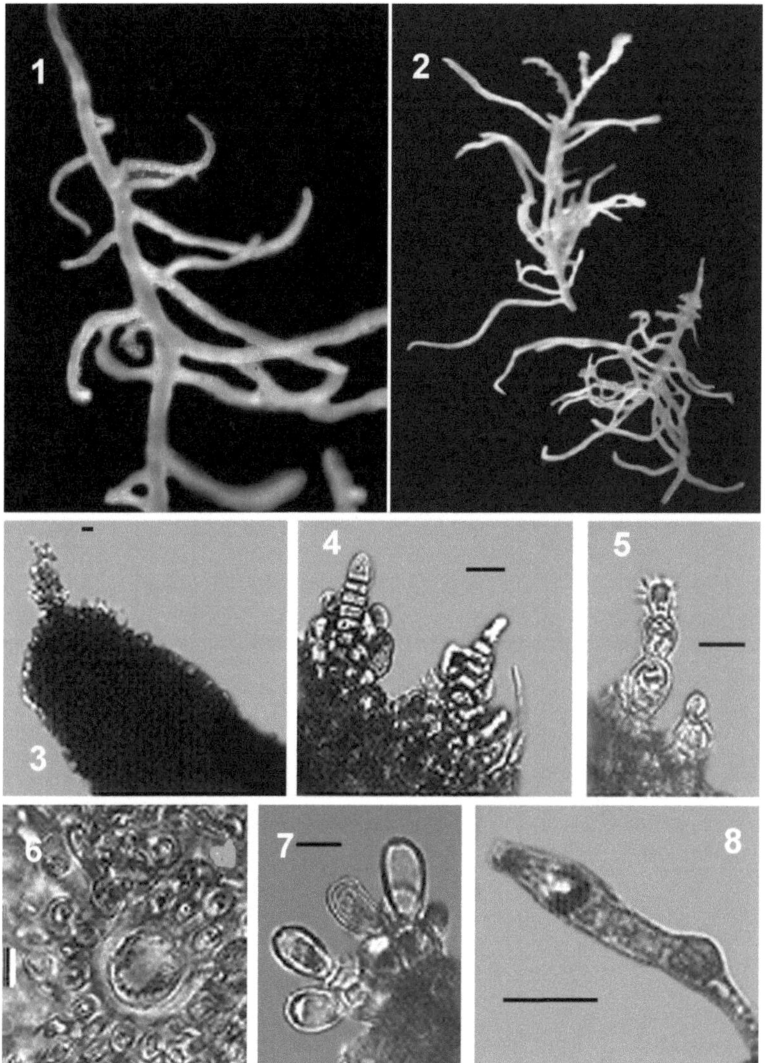

Plate 43: **Fig. 1–8.** *Tuomeya americana*: **1–2.** habitus of plants showing typical ramification and a dense structure with invisible whorls; **3.** juvenile gametophyte developing from a basal mass of cells; **4.** apices of two arising lateral branches; **5.** spermatangia terminally located on fascicle; **6.** cross-section of thallus showing axial cell and cortical filaments; **7.** carposporangia; **8**. carpogonium with stalked trichogyne. Scale bars in figures 3–8 = 10 µm.

Important references: Borge (1928); Webster (1958); Starmach (1977); Necchi Jr. & Entwisle (1990); Kaczmarczyk et al. (1992); Kwandrans & Eloranta (1994); Eloranta & Kwandrans (2002, 2007); Kumano (2002).

Tuomeya americana (Kützing 1857) Papenfuss 1958 (plate **37**: fig. **2**; plate **42**: fig. **1–5**; plate **43**: fig. **1–8)**

Baileya americana Kützing 1857, *Tuomeya fluviatilis* (Harvey) Harvey 1858

Plants, monoecious, bushy, 2.5–5 cm high, ca. 2 mm in diameter (reaching 5 mm in diameter at their base). Branching irregular. Main axis covered with compact cortical filaments of parenchymatous type. No distinct outer cortex. Whorls abundantly branched at the ends, their branchings contact closely with neighbouring whorls. Spermatangia small, ovoid, 1 or 2 located terminally on lateral branches, 3.5–4.0(–5.5–6.0) µm long. Carpogonium-bearing branches short, arising from periaxial cells and or cells of fascicles. Carpogonia 38–43 µm long, asymmetrical with elongated conical or clavate trichogynes, 9–10 µm in diameter. Carposporophyte irregularly compact. Carposporangia growing radially from a dense mass of cells and are terminating on the gonimoblast filaments, 9–10 µm in diameter, 15–16 µm long.
Occurrence: In Europe at 4 locations in central and northern Finland. Also in North America, South Africa and China.

2. Family **Lemaneaceae** C. Agardh 1828

Plants monoecious, pseudoparenchymatous, tubular tufts with regularly placed swellings (nodes); carposporophytes developed within the tubular thallus, spermatangia in form of papillae produced at nodes.

Important references: Atkinson (1890, 1904, 1931); Israelson (1942); Magne (1967a, b); Starmach (1977); Vis & Sheath (1992); Sheath et al. (1996a); Kumano (2002).

Key for genera:

1a Axial cells without cortical filaments; ray cells (periaxial cells) T- or L-shaped, closely connected to the outer fascicles; spermatangial papillae in patches on nodes; plants stalked or unstalked, *chantransia*-stage only transitory . . . . **1. Lemanea**

1b Axial cell with cortical filaments; ray cells simple, not abutting upon the outer fascicles; spermatangial papillae located in rings around the node; plants unstalked; *chantransia*-stage permanent . . . . **2. Paralemanea**

1. **Lemanea** Bory 1808

Lemanea subgenus *Sacheria* (Sirodot 1872) Atkinson 1931, *Sacheria* Sirodot 1872
Plants consisting of tufts of tubular bristle-like axes with regularly placed swellings (nodes). Each axis with an axial filament of very long cells, which are naked without cortical cell filaments. Each axial cell having four T- or L-shaped ray cells (periaxial cells) representing the basal cell of a fascicle perpendicular to it.

Key for species:

1a Plants unstalked, seldom branched, diameter <0.4 mm **2**

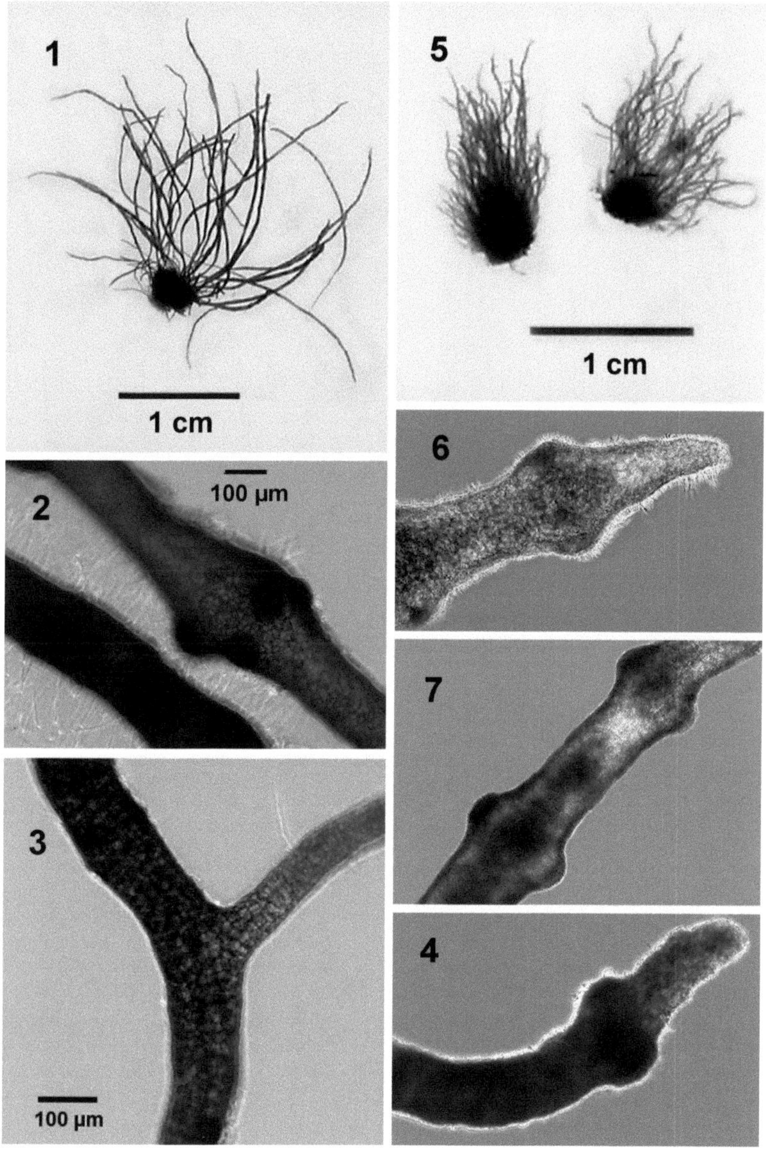

Plate 44: **Fig. 1–4.** *Lemanea borealis*: **1.** tuft of gradually narrowing thalli; **2.** node with series of spermatangial papillae; **3.** branching; **4.** apex of the thallus. **Fig. 5–7:** *Lemanea condensata*: **5.** habitus of tufts; **6.** apex of thallus; **7.** two nodes with spermatangial papillae.

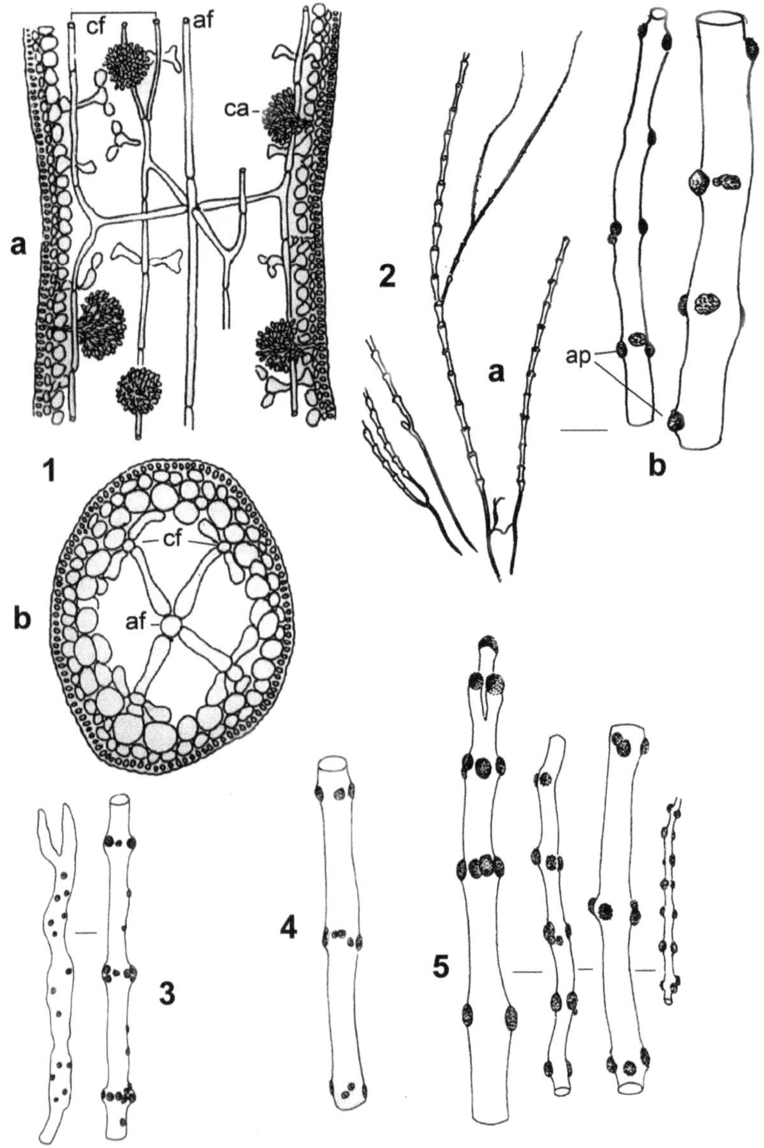

Plate 45: Fig. 1. Structure of *Lemanea* thallus (original author, Starmach 1977): **a** – a longitudinal section of thallus: af = axial filament, cf = cortical filaments, ca = carposporophyte; **b** – a cross-section of thallus. **Fig. 2.** *Lemanea fluviatilis* (original author, Sirodot 1872): **a** – macro-

1b Plants stalked, sparsely to rich branched, diameter >0.3 mm **3**
2a Spermatangial papillae patch like, not pronounced, sterile basal part relatively long . . . . **1. L. borealis**
2b Spermatangial papillae distinctly pronounced, sterile basal part relatively short . . . . **2. L. condensata**
3a Plants narrowing abruptly towards the base (stalked); if branched, usually ≤ 4 branches per plant . . . . **3. L. fluviatilis**
3b Plants narrowing gradually towards the base **4**
4a Spermatangial papillae are distributed irregularly on the branches with indistinct nodes . . . . **4. L. sudetica**
4b Spermatangial papillae are distributed ± regularly, node distinct or indistinct **5**
5a Plants dark violet, spermatangial papillae 2–8 in number, node indistinct . . . . **5. L. ciliata**
5b Plants light or dark green, brownish or yellowish **6**
6a Plants abundantly branched especially at lower parts **8**
6b Plants sparsely branched **7**
7a Plants thick, often curved, olive-green or yellowish spermatangial papillae not protruding . . . . **6. L. rigida**
7b Plants bright or dark green, do not form compact tufts . . . . **7. L. fucina**
8 Plants bright or dark green, forming compact tufts, branches abundant . . . . **8. L. mamillosa**

1. **Lemanea borealis** Atkinson 1904 (plate **44**: fig. **1–4**)

Plants evenly tufted, slender, little branched, 130–350 µm in diameter, 0.6–5.0 cm long; the sterile basal part appears relatively long, 0.5–1.0 cm, slender, unstalked, gradually tapering into the fertile part, the transition is rarely abrupt. Spermatangial zone prominently tuberculate with 2–5 papillae when young. The papillae disappear by age, therefore old plants are plane. Carposporophytes arising from both the spermatangial zone and the carposporophyte zone. Carpospores in tufts throughout the entire length of the plant, ellipsoidal to elongated, 18–25 µm in diameter, 25–45 µm long.
Occurrence: Seems to occur only in northern regions. Europe: Finland, Sweden.

Note: For the discussion of synonyms see Kumano (2002).

2. **Lemanea condensata** Israelson 1942 (plate **44**: fig. **5–7**)

Plants greenish, brownish, blackish- to brownish-olive, unbranched, curved, gradually tapering towards the base. Grows in bunches or forming loose tufts, 1–2(–3) cm in height, in the middle parts 100–250 µm in diameter, internodal zone concave. Spermatangial papillae usually 3, distributed more or less regularly, distinct-

---

scopic view of the thallus; **b** – part of thallus showing spermatangial (antheridial) papillae (ap). **Fig. 3.** *Lemanea sudetica* (original author, Kützing 1857): part of the thallus showing spermatangial papillae. **Fig. 4.** *Lemanea ciliata* (original author, Sirodot 1872): part of the thallus showing spermatangial papillae. **Fig. 5.** *Lemanea rigida* (original author, Sirodot 1872): part of the thallus showing spermatangial papillae.

ly pronounced. Spermatangial papillae zone 220–350 µm in diameter, the papillae included. Carpogonial branches developed along the entire carposporophyte zone and the spermatangial papillae zone. Carposporophyte zone hour-glass-shaped or towards the proximal parts of plants cylindrical, 100–280 µm in diameter in the mid of the plants. Internodes 320–830 µm long, 1.5–3.5 times longer than the diameter in the central part.
Occurrence: Europe: In northern parts of Nordic countries, Finland, Norway and Sweden.

Note: According to Israelson (1942), *L. condensata* has many features common with *L. hispanica* Budde 1929, but he stated that due to a lack of proper diagnosis a closer comparison was not possible.

3. Lemanea fluviatilis (Linnaeus 1753) C. Agardh 1811 (plate **45**: fig. **2**; plate **46**: fig. **1–4**)

Sacheria fluviatilis (Linnaeus) Sirodot 1872, *Lemanea fucina* (Bory) Atkinson var. *subtilis* Atkinson 1890

Plants dark violet or dark brown or yellowish olive, after drying becoming black; thallus narrows abruptly towards the base and transforms into a thin cylindrical stalk. Thallus 6–30 cm long, branching sparse and nodes fairly distinct. Spermatangial papillae usually 3–4, more or less prominent, in mature filaments less so, well separated, occasionally more irregular and partly confluent. Spermatangial zone with papillae 300–1200(–1600) µm in diameter. Carpogonial branches develope all over the generative filaments. Carposporophyte zones cylindrical, initially slightly thinner than spermatangial zones, at maturity slightly broader in the center, 250–750(–1100) µm in diameter. Internodes 1.0–3.0(–4.5) mm long, 3.3–8.0 times longer than the diameter of the central zone of the internode (Israelson 1942).
Occurrence: On stones in rivers. Known from many European countries all over Europe.

4. Lemanea sudetica Kützing 1857 (plate **45**: fig. **3**)

Lemanea kalchenbenneri Rabenhorst 1863?, *Lemanea daldinii* Rabenhorst 1863?

Plants 2–9 cm high, ca. 1 mm in diameter, red violet, brownish, olive-green to dark green, gradually narrowing towards the base. Spermatangial papillae are irregularly distributed over the thallus with not very pronounced nodes (Starmach 1977).
Occurrence: Known from Sudeten Mts. in Poland and Czech Republic, a questionable finding from North Italy is also reported.

Note: In the checklist of non-vascular plants of Slovakia *Lemanea sudetica* is named as a synonym for *Lemanea torulosa* (Marhold & Hindák 1998). However, according to Kumano (2002), *Lemanea torulosa* Kützing should be treated as a synonym of *Lemanea rigida*.

5. Lemanea ciliata (Sirodot 1872) De Toni 1897 (plate **45**: fig. **4**)

Sacheria ciliata Sirodot 1872

Plants dark violet, brownish, olive green or dark green, branching abundant; gradually tapering towards the base, thus the stalk is not very pronounced. Branches ter-

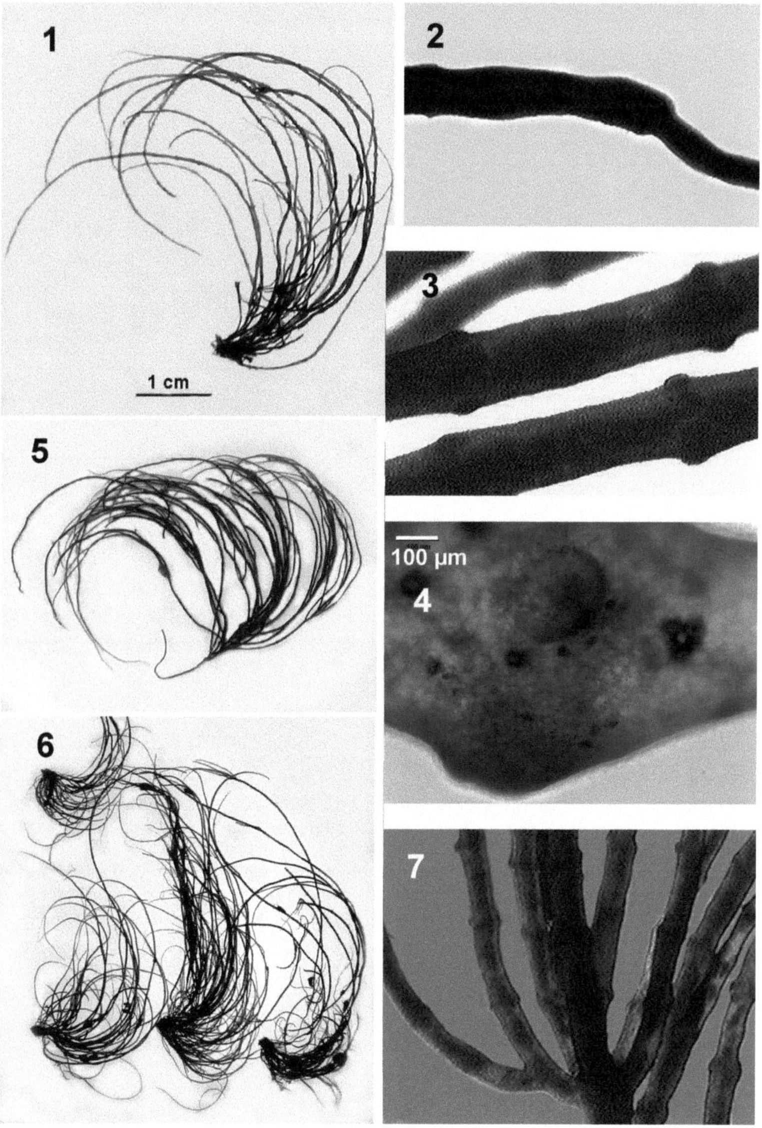

Plate 46: **Fig. 1–4.** *Lemanea fluviatilis*: **1.** habitus of tuft; **2.** abruptly narrowing stalked basal part of the thallus; **3.** part of the fertile zone with two nodes and spermatangial papillae; **4.** – node with spermatangial papillae. **Fig. 5–7.** *Lemanea mamillosa*: **5–6.** habitus of thallus showing rich ramification; **7.** abundant branches of the lower part of the thallus.

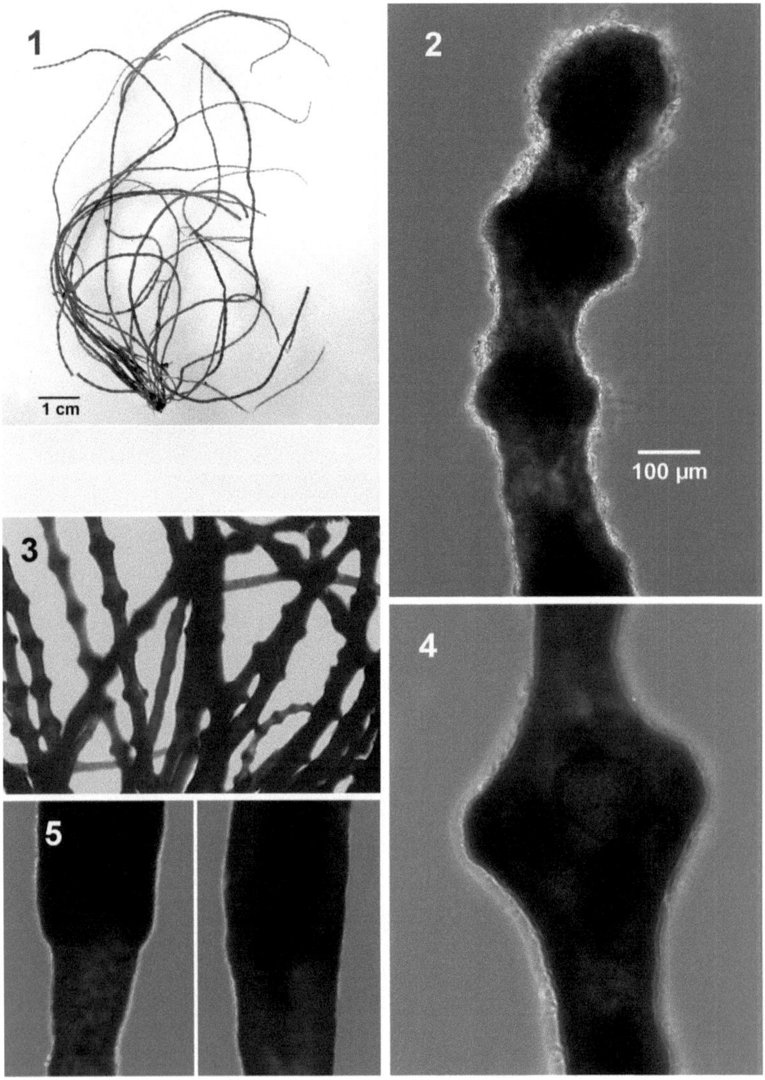

Plate 47: **Fig. 1–5.** *Lemanea rigida*: **1.** habitus of thallus; **2.** apex of thallus with spermatangial papillae; **3.** thallus showing clearly concave internodal zones; **4.** node with spermatangial papillae; **5.** two types of basal parts of the fertile zone tapering to the sterile stalk.

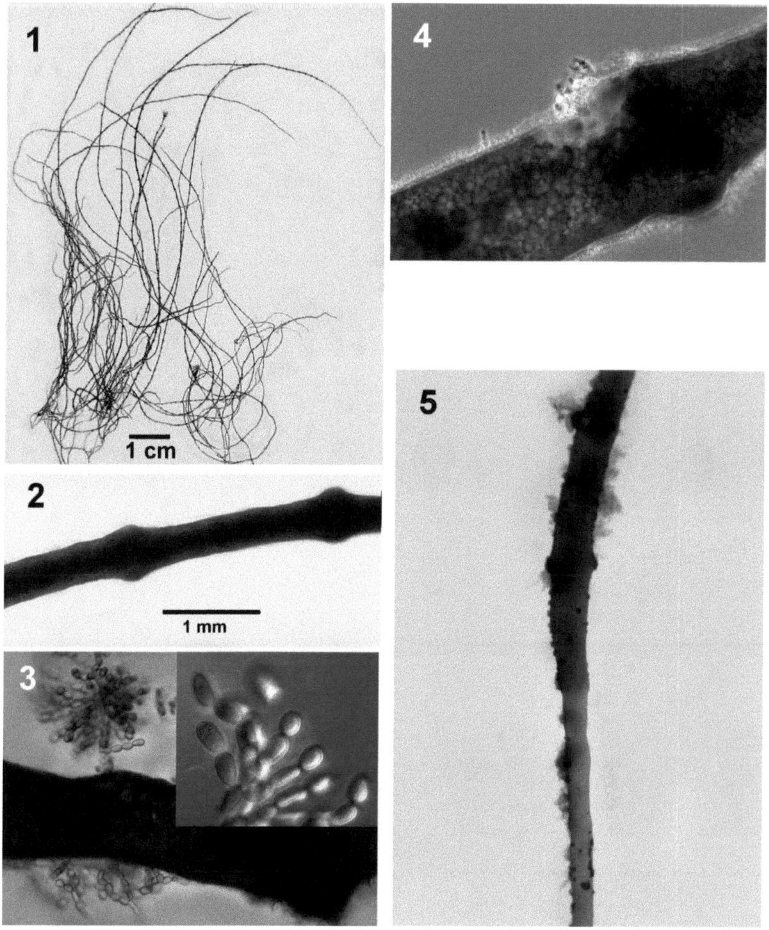

Plate 48: **Fig. 1–5.** *Lemanea fucina*: **1.** habitus of thallus; **2.** part of thallus with two nodes showing spermatangial papillae; **3.** broken thallus with carposporophyte and carpospores terminal on gonimoblast filaments (inserted photo); **4.** thallus showing cortical structure and spermatangial papillae; **5.** lower part of thallus showing gradually tapering area with small epiphytic tufts of *Audouinella hermannii*.

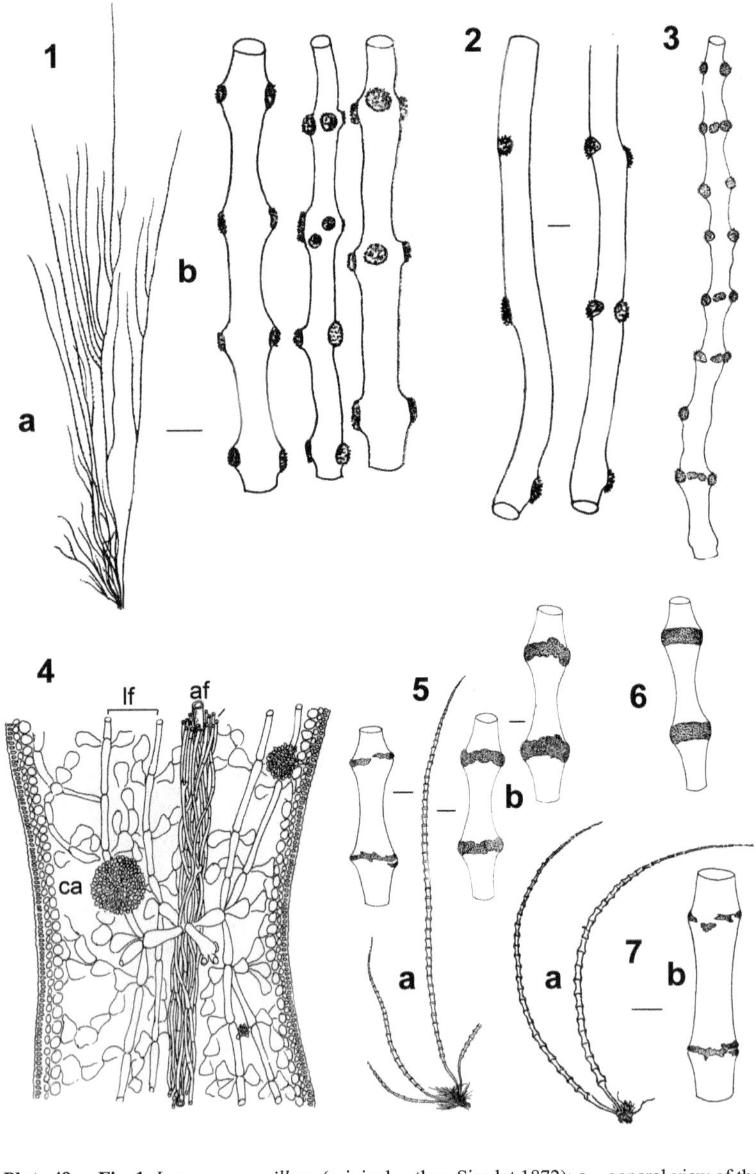

Plate 49: **Fig. 1.** *Lemanea mamillosa* (original author, Sirodot 1872): **a** – general view of the plant showing rich branching; **b** – parts of thallus showing spermatangial papillae. **Fig. 2.** *Lemanea fucina* (original author, Sirodot 1872): parts of thallus showing spermatangial papillae.

minating in terminal hairs. Spermatangial papillae are distributed more or less regularly, not very pronounced, 3–8 per node.
Occurrence: Known from Ille-Rance Canal in Bretagne, France and from Poland.

6. Lemanea rigida (Sirodot 1872) De Toni 1897 (plate **45**: fig. **5**; plate **47**: **1–5**)

Lemanea dichotoma De Candolle 1805, *Lemanea fucina* Bory var. *rigida* Atkinson 1890, *Lemanea torulosa* sensu Kützing 1843 (not *Lemanea torulosa* (Roth 1797) C. Agardh (1814)

Plants thick, rough and curved, 5–20 cm long, branching at the base, gradually tapering towards the base. Colour red-violet, brownish or olive-green. Spermatangial papillae 3–7 in a whorl, more or less regularly distributed, nearly to completely plane, often confluent, sometimes protruding but not fusing into rings. Old plants are becoming yellowish during desiccation.
Occurrence: In running waters; Europe: Estonia, Belgium, Finland, France, Slovakia, Sweden.

7. Lemanea fucina Bory 1808 (plate **48**: **1–7**; plate **49**: fig. **2**)

Lemanea mamillosa var. *fucina* Kützing 1843, *Sacheria fucina* (Bory) Sirodot 1872, *Sacheria mamillosa* Sirodot 1872

Young plants olivaceous, or yellowish-green, sometimes dark violet, usually preserving colour during desiccation. Old plants become yellowish or greenish when drying; size 2–40 cm, very delicate or stout, not in compact tufts. Spermatangial papillae 2–7 (usually 3), not very pronounced, internodal zone cylindrical or incised.
Occurrence: In fast flowing creeks and streams; Europe: France, Germany and northern Europe (Finland, Norway, Sweden). Records from British Isles are given as a synonym to *L. mamillosa* and *L. rigida* (Sheath & Sherwood 2002).

8. Lemanea mamillosa Kützing 1845 (plate **46**: fig. **5–7**; plate **49**: fig. **1**)

Sacheria mamillosa (Kützing 1845) Sirodot 1872, *Lemanea fucina* var. *mamillosa* (Kützing 1845) Atkinson 1890

Plants forming tufts, stout, 4–20 cm high, branches abundant, dark green, brownish or olive green; at the basal part only slightly stalked. Spermatangial papillae 3–4 in whorls, more or less regularly distributed.
Occurrence: In rivers and creeks, tolerant to rather strong eutrophication; Europe: Estonia, Finland, France, Germany and Sweden. Probably also reported from British Isles under the name *L. fucina* (Sheath & Sherwood 2002).

Fig. 3. *Lemanea parvula* (original author, Sirodot 1872): part of thallus showing spermatangial papillae. **Fig. 4.** Structure of *Paralemanea* thallus (original author, Starmach 1977): longitudinal section of thallus: af = axial filament surrounded by cortical filaments, lf = lateral filaments, ca = carposporophyte. **Fig. 5.** *Paralemanea catenata* (original author, Sirodot 1872): **a** – general view of the thallus; **b** – parts of thallus showing spermatangial papillae. **Fig. 6.** *Paralemanea annulata* (original author, Sirodot 1872): part of thallus showing spermatangial papillae. **Fig. 7.** *Paralemanea torulosa* (original author, Sirodot 1872): **a** – general view of the plant; **b** – part of thallus showing spermatangial papillae.

Uncertain species:

Lemanea parvula Sirodot 1872 (plate **49**: fig. **3**). This species is placed in Starmach (1977) in subgenus *Paralemanea* due to coalescent spermatangial papillae. However, as papillae do not form disticnt rings, this taxon should be kept in *Lemanea*. Since the original description done by Sirodot (1872), this species was not found again; it is not included in Kumano (2002).
Plants form small tuft with radial branches from the base. Spermatangial papillae form separate or sometimes coalescent tubercles: thallus 6–7 cm high, olive-green or yellow.
Occurrence: Known from rivers in France.

2. **Paralemanea** (Silva 1959) Vis et Sheath 1992

Lemanea subgenus *Paralemanea* Silva 1959, *Lemanea* Sirodot 1872

Axial cell with cortical filaments; ray cells simple, not abutting the outer fascicles; spermatangial papillae arranged in rings around the nodes, sometimes scattered; plants unstalked and usually unbranched; *chantransia*-stage permanent.

Important references: Starmach (1977); Vis & Sheath (1992); Sheath et al. (1996a); Kumano (2002); Kučera & Marvan (2004)

Key to the species:

1a	Plants ≥8 cm high, ≥0.7 mm in diameter, internodia concave	**2**
1b	Plants <6 cm long, <0.7 mm in diameter, internodia ± parallel	**1. P. torulosa**
2a	Spermatangial rings narrow, irregular	**2. P. catenata**
2b	Spermatangial rings broad, regular	**3. P. annulata**

1. **Paralemanea torulosa** (Roth 1797) Sheath et Sherwood 2002 (plate **49**: fig. **7**; plate **50: 1–6;** plate **51**: **1–6)**

Conferva torulosa Roth 1797, *Lemanea torulosa* (Roth) C. Agardh 1814

Plants 5–8(–15) cm long, violet or olive-green, unstalked and unbranched (according to Starmach 1997 sometimes branched), bristle-like 0.5–0.7(–1.0) mm in diameter. Spermatangial zones form narrow broken rings at regular intervals. Internodia almost parallel or slightly incissed.
Occurrence: In rivers; Europe: Belgium, British Isles, Czech Republic, France, Germany, Switzerland.

Note: This species is reported from one locality in Czech Republic by Kučera & Marvan (2004) as *Lemanea torulosa* (Roth 1797) C. Agardh 1814. However, from their description the identificiation remains unclear.

2. **Paralemanea catenata** (Kützing 1843) Vis et Sheath 1992 (plate **49**: fig. **5**)

Lemanea catenata Kützing 1843, *Lemanea nodosa* Kützing 1867, *Lemanea pleocarpa* Atkinson 1931

Plants (5–)8–30 cm long, violet or olive-green, unstalked and unbranched, 0.7–0.9(–1.5) mm in diameter, distinctly narrowing at their base, straight or slightly arcuate. Grows in tufts. Spermatangial papillae narrow, forming irregular rings, sometimes interrupted near the base. Carposporophyte zones concave, carpogonial branches arising from the center of the carposporophytes zone.
Occurrence: In rivers; Europe: Belgium, France, Germany and Spain.

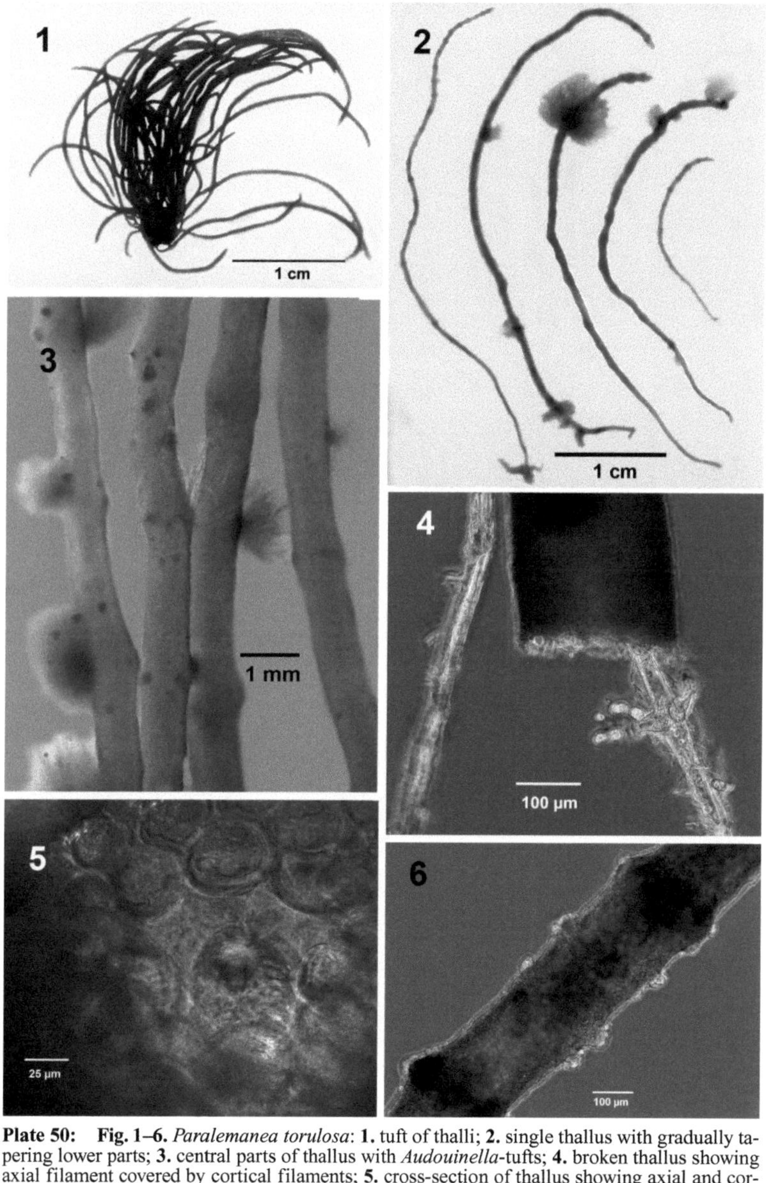

Plate 50: **Fig. 1–6.** *Paralemanea torulosa*: **1.** tuft of thalli; **2.** single thallus with gradually tapering lower parts; **3.** central parts of thallus with *Audouinella*-tufts; **4.** broken thallus showing axial filament covered by cortical filaments; **5.** cross-section of thallus showing axial and cortical filaments in the center; **6.** part of thallus with broken rings of spermatangial papillae and an almost cylindrical internode.

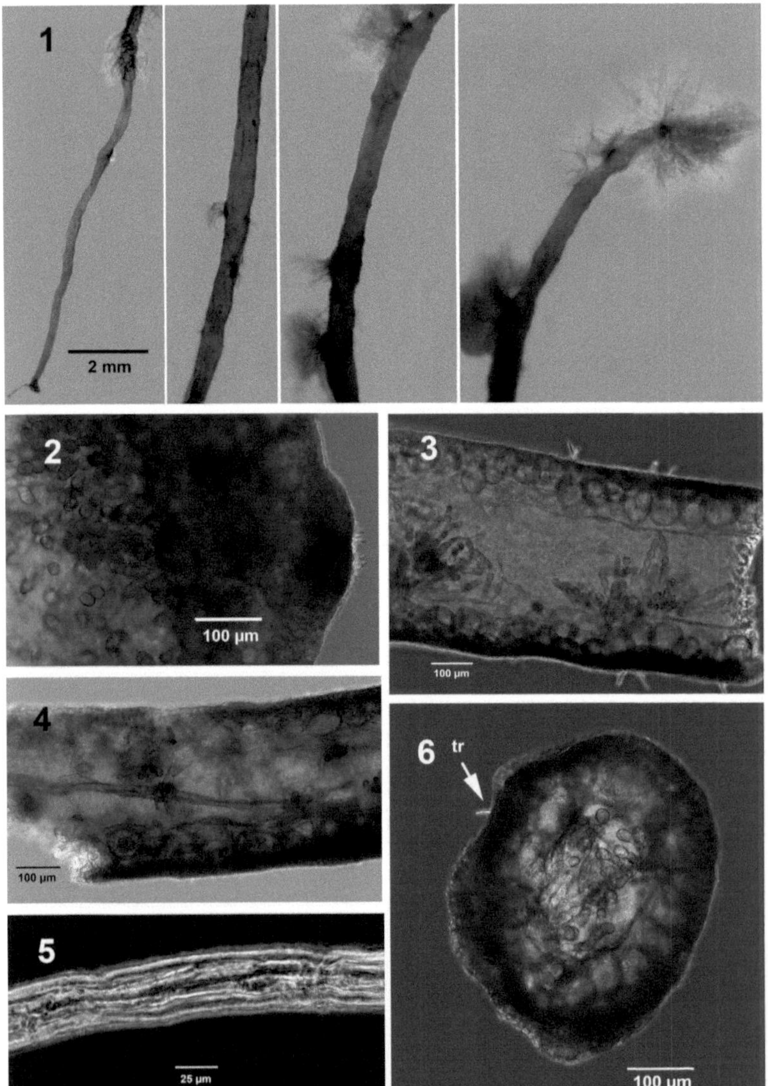

Plate 51: Fig. 1–6. *Paralemanea torulosa*: **1.** parts of a single thallus from the gradually tapering basal part towards the apex; **2.** longitudinal section of thallus with carpospores inside the tubular thallus and spermatangial papilla to the right; **3.** longitudinal section of thallus showing cortex layers and carposporophytes; **4.** longitudinal section of thallus with central axial filament; **5.** axial filament with cortical filaments on the surface; **6.** cross-section of the thallus (tr = trichogyne).

3. Paralemanea annulata (Kützing 1843) Vis et Sheath 1992 (plate **49**: fig. **6**; plate **52**: **1–5**)

Lemanea annulata Kützing 1843

Plants 8–15 cm long, violet or dark brown, unstalked and very rarely branched, in nodes up to 2 mm in diameter, concave at internodes. Spermatangial papillae form regular, broad rings, sometimes interrupted near the base. Rings at the time of fertilization brighter in colour than the other parts of thallus. Spermatangial zones sometimes enlarging after fertilization by increased growth of the tissue at the base of the spermatangia. Carposporophytes developing in the center of the carposporophyte zone. Occurrence: In rivers; Europe: France, Germany, Switzerland.

Note: Kučera & Marvan (2004) included *Paralemanea annulata* to *P. catenata* in their contribution on Lemaneaceae of Czech Republic.

8. Order **Thoreales** Müller, Sherwood, Pueschel, Gutell et Sheath 2002

Thallus mucilaginous, densely branched, multiaxial filaments with lateral assimilating filaments.

Important references: Starmach (1977); Necchi Jr. (1987); Sheath et al. (1993b, 2000); Necchi Jr. & Zucchi (1997); Kumano (2002); Müller et al. (2002)

Note: Reproductive organs, which earlier have been reported as monosporangia for some taxa may represent either spermatangia (smaller ones) or carposporangia (larger ones) (Necchi Jr. 1987).

Family **Thoreaceae** (Reichenbach) Hassall 1845

Thallus abundantly branched, densely covered with hairs, which are formed by assimilating filaments situated perpendicularly to the main longitudinal axis. Two genera, *Thorea* and *Nemalionopsis*, but only *Thorea* occurs in Europe.

Thorea Bory 1808

Plants gelatinous, abundantly branched, multiaxial, the central part is called medulla, lateral assimilatory filaments disposed along the main axis.

Important references: Starmach (1977); Necchi Jr. (1987); Sheath et al. (1993b, 2000); Lokhorst (1994); Necchi Jr. & Zucchi (1997); Kumano (2002); Müller et al. (2002)

Key to the species:

1a Plants up to 100 cm or more in height, 500–1890 µm wide, assimilatory filaments 700–1400 µm long .... **1. T. hispida**

1b Plants up to 50 cm high, central axis >2000 µm in diameter, assimilatory filaments 750–1415 µm long .... **2. T. violacea**

1. Thorea hispida (Thore) Desvaux 1818 emend. Sheath, Vis et Cole 1993 (plate **53**: fig. **1**; plate **54**: fig. **5–7**)

Conferva hispida Thore 1799, *Conferva flexuosa* Bory 1804 nom.illeg., *Thorea ramosissima* Bory 1808 illeg., *Thorea lehmannii* Hornemann 1818, *Thorea andina* Lagerheim et Möbius in Möbius 1891

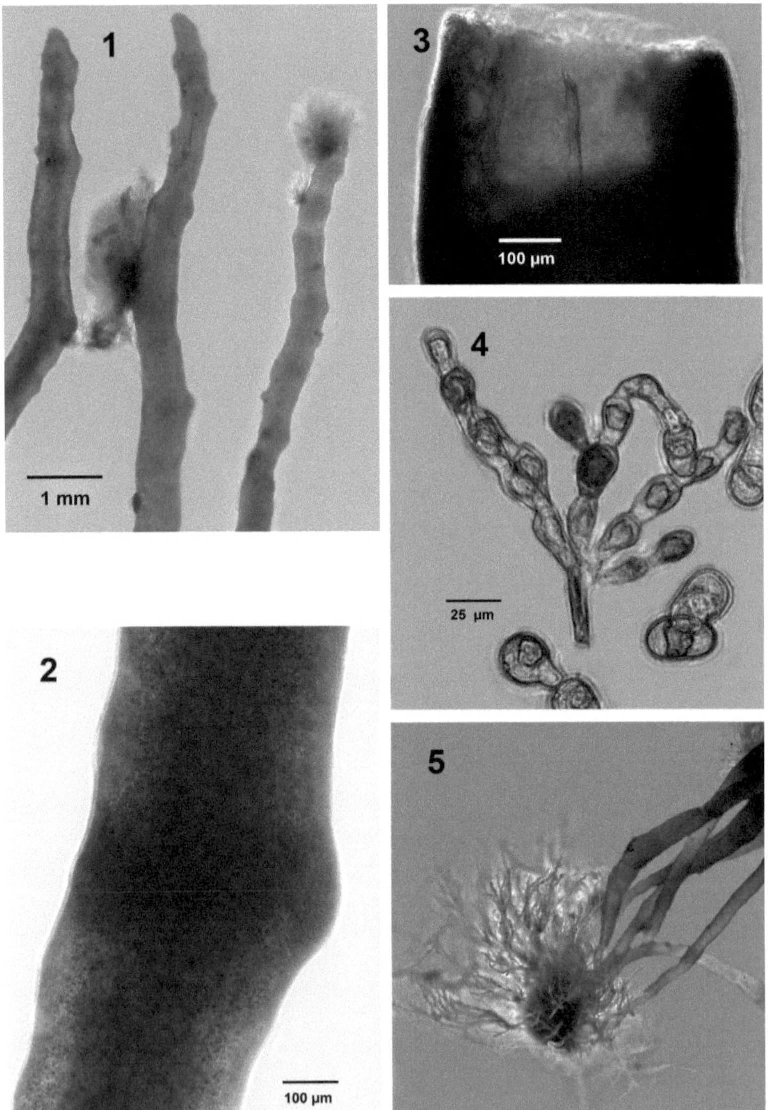

Plate 52: Fig. 1–5. *Paralemanea annulata*: **1.** apices of thalli with tufts of *Audouinellas*; **2.** regular ring of spermatangial papillae; **3.** section of thallus with axial filament covered by thin cortical filaments; **4.** gonimoblast filament and carpospores; **5.** basal parts of thalli and filaments of permanent *chantransia*-stage.

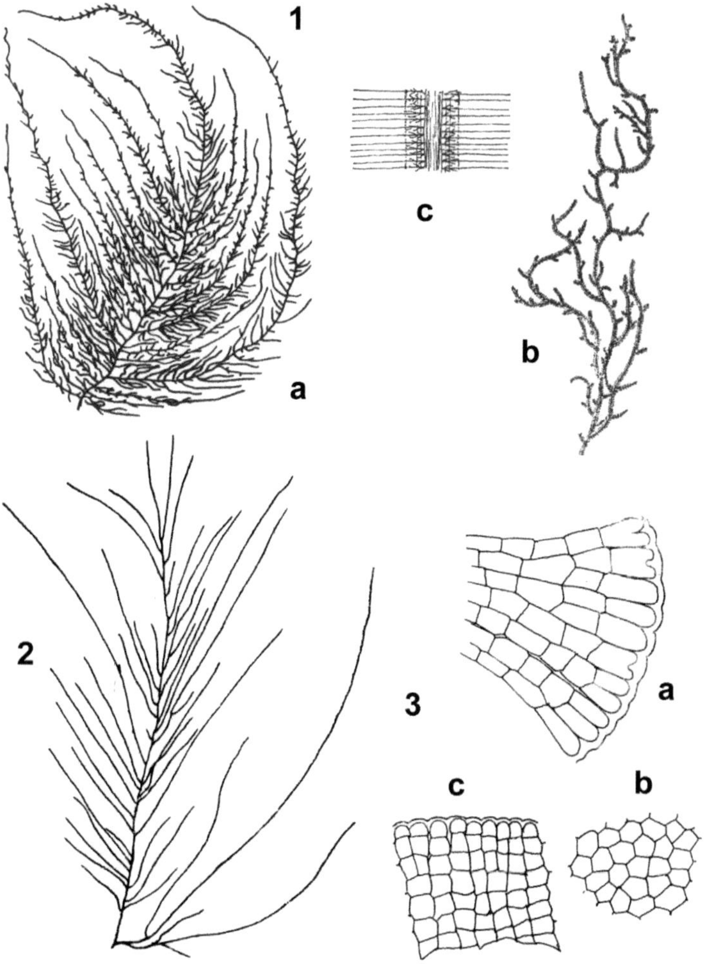

Plate 53: **Fig. 1.** *Thorea hispida* (original authors, Klas 1936: a, Starmach 1977: b, c): **a, b** – general view of thallus with abundant branching; **c** – part of thallus showing central medulla and thin unbranched assimilatory filaments. **Fig. 2.** *Thorea violacea* (original author, Klas 1936): general view of the thallus showing branching. **Fig. 3.** *Hildenbrandia rivularis* (original author, Starmach 1977): **a** – view of patch fragment showing elongated basal layer cells; **b** – cells of erect filaments (view of thallus surface); **c** – dense erect filaments.

Plants ≥100 cm high, abundantly branched, secondary branches 3 to 14 per cm, plant diameter 513–1890 µm, main axis 88–611 µm in diameter. Assimilatory filaments sparsely branched, 700–1400 µm long. Chloroplast laminate, greenish or violet. A mucilaginous envelope surrounds the axial part of the plant covering the basal parts of the assimilatory filaments. Sporangia relatively rare, 8.6–30.0 µm long, in clusters up to 8 sporangia.
Occurrence: In rivers and ditches; rarely found. Europe: Austria, Belgium, British Isles, Croatia, France, Germany, Hungary, Lithuania, The Netherlands, Poland, Rumania, Serbia, Slovakia and Spain.

2. Thorea violacea Bory de Saint-Vincent 1808 (plate **53**: fig. **2**)

Thorea brodensis Klas 1935

Plants 20–50 cm high, robust, whip-like, unbranched, sometimes sparsely branched, central axis ≥2000 µm in diameter, main axis evident without assimilatory filaments. Secondary branches ca. 200 µm in diameter, with rich-branched assimilatory filaments 750–1415 µm long. Sporangia ellipsoid or pear-shaped, 13–20 µm long, either single on a small stalk or in clusters on fascicles, inserted alternately or in opposite on main axis or on secondary axes.
Occurrence: Known from Sava River near Slavonski Brod, Croatia and the saline spring of Marjal Pego-Oliva, Spain.

Note: Sheath et al. (1993b) regarded *Thorea brodensis* as conspecific with *Thorea violacea* Bory 1808.

9. Order **Hildenbrandiales** Pueschel et Cole 1982

Thallus crustaceous forming a basal layer developed from dense, laterally aggregated filaments. Each cell produces an erect filament upwards. Sporangium-bearing branches short, sparsely branched and assimilatory filaments loosely arranged. Sexual reproduction unknown.

Important references: Skuja (1926b); Starmach (1977); Sheath et al. (1993b, 2000); Sherwood & Sheath (1999, 2000, 2003); Kumano (2002); Sherwood et al. (2002)

Family **Hildenbrandiaceae** Rosenvinge 1917

See description for the order. One genus described.

Hildenbrandia Nardo 1834 in NUC-3 1993 nom. et orth. cons.

Plants crustaceous, forming red patches on stones in shaded streams and in lakes. Vegetative reproduction by fragments of plants, and by means of offshoots and gemmae.

Key to the species:

1a Cell diameter 4–12 µm (mean 8.2 µm), cell length 9–13 µm (mean 11.3 µm) . . . . **1. H. rivularis**

1b Cell diameter 4–7 µm (mean 4.9 µm), cell length 5–9 µm (mean 6.4 µm) . . . . **2. H. angolensis**

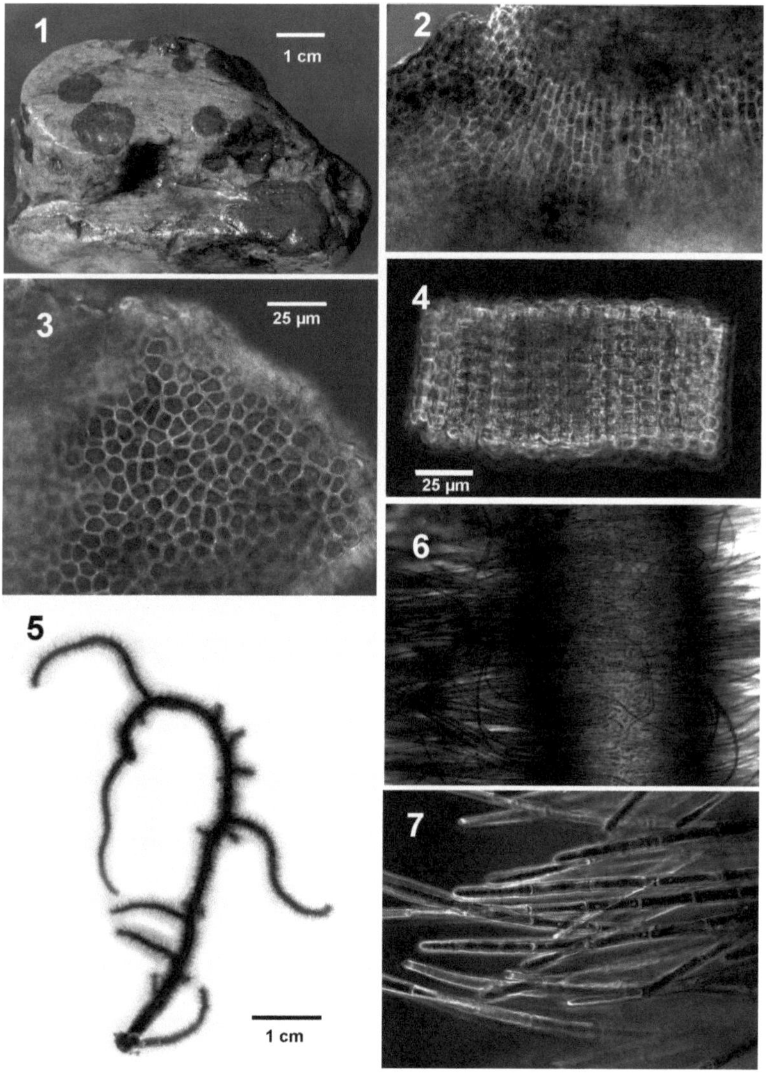

Plate 54: **Fig. 1–4.** *Hildenbrandia rivularis*: **1.** round crustaceous patches of plant on the surface of a stone; **2–3.** surface views of thallus; **4.** side view of filaments (erect filaments). **Fig. 5–7.** *Thorea hispida*: **5.** habitus of the plant showing ramification, dense medulla and thin assimilatory filaments; **6.** details of a central part of the thallus; **7.** assimilatory filaments.

Note: All measures of the species are overlapping, which indicates that *H. angolensis* is conspecific with *H. rivularis*. However, according to the results of the molecular analyses by Sherwood & Sheath (2000), those taxa should be kept separately.

1. Hildenbrandia rivularis (Liebmann) J. Agardh 1851 (plate **53**: fig. **3**; plate **54**: fig. **1–4**)

Erythroclathrus rivularis Liebmann 1839

Plants bright crimson, forming more or less regularly circular discs, with diameter up to 2.5 cm; neighbouring discs often fusing into extensive crusts over the stone. Discs about 50–100 µm thick, consisting of a basal layer composed of radiating rows of somewhat elongated cells and erect filaments. Erect filaments densely aggregated, simple or rarely dichotomously branched and composed of almost isodiametric cells, usually 7–10(–15) cells per filament, 4–12 µm in diameter. Asexual propagation by forming of gemmae, and fragmentation of erect filaments.
Occurrence: In shaded rivers and in lakes; favours waters with higher conductivity (e.g. calcareous waters); known from many European countries from the North to the South.

2. Hildenbrandia angolensis Welwitsch ex W. West et G.S. West 1897 (no figures)

Like *H. rivularis*; discs usually 21–162 µm thick (4–17 cell layers), cells 4–7 µm in diameter, 5–9 µm long.
Occurrence: In Europe known only from Spain; other records are from tropical and subtropical areas.

References to Rhodophyta

Aboal Sanjurjo M. (1989): Contribución al conocimiento de las algas epicontinentales del SE de España II: Los rodófitos (*Rhodophyceae*). – Lazaroa **11**: 115–122.

Agardh C.A. (1812): Dispositio algarum Suecicae. – Lund, pp. 1–16 (1810), 17–26 (1811), 27–45.

Agardh C.A. (1814): Anmärkingar om slägtet *Lemanea*, samt beskrivning om tvenne nya arter deraf. – Kongl. Sv. Vetensk.-Akad. Handl. **35**: pp. 33.

Agardh C.A. (1824): Systema algarum. – Lund, pp. xxxviii + 312.

Agardh C.A. (1928): Species algarum, Vol. 2, sect. 1. – Greifswald, pp. lxxvi + 189.

Agardh J.G. (1847): Nya alger från Mexico. – Öfvers. Förh. Kongl. Sv. Vetensk.-Akad. **4**: 5–17.

Agardh J.G. (1851): Species, genera et ordines algarum (Fucoidarum). Vol. 2. – Lund; part 1: I–XII + 1–336, part 2, fasc. 1: 337–504.

Albertano P., Ciniglia C., Pinto G. & A. Pollio A. (2000): The taxonomic position of *Cyanidium*, *Cyanidioschyzon* and *Galdieria*: an update. – Hydrobiologia **433**: 137–143.

Alvarez Cobelas M. (1984): Catálogo de las algas continentals españolas. II. Craspedophyceae, Cryptophyceae, Chrysophyceae, Dinophyceae, Euglenophyceae, Haptophyceae, Phaeophyceae, Rhodophyceae, Xanthophyceae. – Acta Bot. Malacitana **9**: 27–40.

Atkinson G.F. (1890): Monograph of the Lemaneaceae of the United States. – Annals of Botany **4**(14): 177–229 + 3 tables.
Atkinson G.F. (1904): A new *Lemanea* from Newfoundland. – Torreya **4**: 26.
Atkinson G.F. (1931): Notes on the genus *Lemanea* in North America. – Bot. Gaz. **92**: 225–242.
Backhaus D. (1973): Fliessgewässeralgen und ihre Verwendbarkeit als Bioindikatoren. – Verh. Ges. Ökol. Saarbrücken **8**: 149–163.
Basson P.W. (1979): Marine algae of the Arabian Gulf coast of Saudi Arabia. – Bot. Mar. **22**: 47–82.
Belcher J.H. & Swale E.M.F. (1960): Some Britisch freshwater material of *Asterocystis*. – Br. Phycol. Bull. **2**(1): 33–35.
Bjørnland T., Borch G. & Liaaen-Jensen S. (1984): Configurational studies on red algal caretoneids. – Phytochemistry **23**: 1711–1715.
Borge O. (1928): Süsswasseralgen: – In: B. Schröder: Zellpflanzen Ostafrikas. Teil VIII. – Hedwigia **68**: 93–114.
Bory de Saint-Vincent J.B. (1797): Memoire sur les genres *Conferva* et *Byssus*, du chevalier O. Linné. – Louis Cavazza, Bordeaux, pp. 58.
Bory de Saint-Vincent J.B. (1804): Voyage dans les quatre principales iles des mers d'Afrique. Tom. 2. – Buisson, Paris, pp. 43.
Bory de Saint-Vincent J.B. (1808a): Mémoire sur le genre *Batrachosperma*, de la famille des Conferves. – Ann. Mus. Hist. Nat. **12**: 310–332.
Bory de Saint-Vincent J.B. (1808b): Sur le genre *Lemanea* de la famille des Conferves. – Ann. Mus. Hist. Nat. **12**: 177–190.
Bory de Saint-Vincent J.B. (1823a): *Audouinella*. – Dict. Class. Hist. Nat. **3**: 340–341
Bory de Saint-Vincent J.B. (1823b): Ceramiales et Confervées. – Dict. Class. Hist. Nat. **3**: 392–394.
Bourrelly P. (1985): Les algues d'eau douce. Initiation à la systematique. Tome III: Les algues bleues et rouges, les Eugleniens, Peridiniens et Cryptomonadines. – 2nd ed., Paris, Èditions N. Bourbée et Cie, pp. 512.
Brand F. (1910): Über die Süsswasserformen von *Chantransia* (DC) Schmitz, einschliesslich *Pseudochantransia* Brand. – Hedwigia **36**: 107–118.
Budde H. (1929): Beitrag zur Algenflora der fliessenden Gewässer Spaniens. – Arch. f. Hydrobiol. **20**: 424–440.
Burkholder J.M. & Sheath R.G. (1985): Characteristics of softwater streams in Rhode Island. I. A comparative analysis of physical and chemical variables. – Hydrobiologia **128**: 97–108.
Cambra J. (1985): Catàlegs de les especies de cianofícies, d'algue continentalis I del plankton marí dels Països Catalans. Annex I. – In: Llimona X. et al. (eds.), Enciclopedia Catalana, Historia Natural des Països Catalans. Plantes inferiors, Vol. **4**: 501–522, Barcelona, pp. 588.
Cambra J. (1987): *Lemanea nodosa Kütz.* 1843 en el torrente de Arbúcies (Barcelona). – Act. VI Simp. Nac. Bot. Criptog.: 19–24.
Chapman D.J. (1974): Taxonomic status of *Cyanidium caldarium* the Porphyridiales and Goniotrichales. – Nova Hedwigia **25**: 673–682.
Christensen T. (1980): Algae, a taxonomic survey. Fasc.1. – Odense, AiO Tryck.
Collins F.S. (1895): A collection of dried specimens of the algae of North America. – Phycotheca Boreali-Americana: 990.
De Candolle A.P. (1801): Extrait d'un rapport sur les Conferves, fait à la Société philomathique. – Bull. Sci. Soc. Philom. Paris **3**: 17–2.

De Candolle A.P. (1805): Flore Francaise, Vol. 6. – Paris.
Dell Grosso F. (1981): Prima segnalazione in Italia del genere Sirodotia con la specie S. huillensis (Welw.) Skuja, nuova per l'Europa. – **115**: 11–16.
De Luca P., Taddei R. & Varano L. (1978): *"Cyanidioschyzon merolae"*: a new alga of thermal acidic environments. – Webbia **33**: 37–44.
De Toni G.B. (1897): Sylloge algarum. Vol. IV. Florideae. Sectio I. – Padova, pp. (I)–XX + (I)–LXI + (1)–388.
den Hartog C. & Lokhorst G.M. (1995): Occurrence of the Rhodophycean alga *Thorea ramosissima* Bory in the river Meuse. – Netherl. J. Aquat. Ecol. **29**(1): 117–120.
Descy J.-P. & Empain A. (1974): *Thorea ramosissima* Bory (Rhodophyceae, Nemalionales) dans le basin Mosan Belge. – Bull. Soc. Roy. Bot. Belg. **107**: 23–26.
Desvaux N.A. (1809): Plantes trouvées dans le Haut-Poitoules unes nouvelles les autres, point, indiquées dans la flore de France. – J. Botan. **2**: 307–318.
Desvaux N.A. (1818): Observations sur les plantes environs d'Angers. – Angers et Paris, pp. 188.
Diner B.A. (1979): Energy transfer from the phycobilisomes to photosystem II reaction centers in wild type Cyanidium caldarium. – Plant Physiol. **63**: 30–34.
Dixon P.S. (1973): Biology of the Rhodophyta. – Edinburgh, Oliver & Boyd, pp. 285.
Dixon P.S. (1982): Rhodophycota. – In: Parker S.P. (ed.), Synopsis and classification of living organisms. Vol.**1**: 61–69, New York, McGraw-Hill.
Drew K.M. (1935): The life history of *Rhodocorton violaceum* (Kütz.) comb. nova, (*Chantransia violacea* Kütz.). – Ann. Bot. **49**: 439–450.
Drew K.M. (1946): Anatomical observations on a new species of *Batrachospermum*. – Ann. Bot. (London), ser. 2, **10**: 339–352.
Drew K.M. & Ross R. (1965): Some generic names in the Bangiophycideae. – Taxon **14**: 93–99.
Dring M.J. (1982): The biology of the marine plants. – London, Edward Arnold, pp. 199.
Duby J.E. (1830): Aug. Pyrami de Candolle Botanicon Gallicum. Pars. 2. Plantas cellulares continens. – Paris, pp. 545–1068 + i–lviii.
Egidos A.I. & Aboal M. (2003): *Thorea violacea* Bory de Saint-Vincent (Thoreaceae, Rhodophyceae) en una surgencia del Marjal Pego-Oliva, Comunidad Valenciana. Nueva cita para la flora algal Española. – Anales Jard. Bot. Madrid **60**(1): 27–32.
Eisen J.A. & Fraser C.M. (2003): Phylogenomics: Intersection of evolution and genomics. – Science **300** (5626): 1706–1707.
El-Gamal A.D. & Salah El-Din R.A. (1999): New species of the genus *Compsopogon* (*C. helwanii*) from Egypt. – Phycos **38**: 37–42.
Eloranta P. & Kwandrans J. (1996): Distribution and ecology of freshwater red algae (Rhodophyta) in some Finnish rivers. – Nordic J. Botany **16**: 107–117.
Eloranta P. & Kwandrans J. (2002): Notes on some interesting freshwater Rhodophyta from Finland. – Algol. Studies **105**: 95–109.
Eloranta P. & Kwandrans J. (2004): Indicator value of freshwater red algae in running waters for water quality assessment. – Oceanol. Hydrob. Stud. **33**(1): 47–54.
Eloranta P. & Kwandrans J. (2007): Freshwater red algae (Rhodophyta). Identification guide to European taxa, particularly to those in Finland. – Norrlinia **15**: 1–103.

Engler H.G. (1892): Syllabus der Vorlesungen über spezielle und medizinische-pharmaceutische Botanik. – Berlin, pp. xxiii + 184.
Engler H.G. & Prantl K.A.E. (1896): Die natürlichen Pflanzenfamilien. I (1, 2). – Leipzig, Engelmann.
Entwisle T.J. & Foard H.J. (1997): *Batrachospermum* (Batrachospermales, Rhodophyta) in Australia and New Zealand: New taxa and emended circumscriptions in sections *Aristata*, *Batrachospermum*, *Turfosa* and *Virescentia*. – Austr. System. Bot. **10**: 331–380.
Entwisle T.J. & Foard H.J. (1999): Freshwater Rhodophyta in Australia: *Ptilothamnion richardsii* (Ceramiales) and *Thorea conturba* sp. nov. (Batrachospermales). – Phycologia **38**(1): 47–53.
Entwisle T.J. & Kraft G.T. (1984): Survey of freshwater red algae (Rhodophyta) of south-eastern Australia. – Austr. J. Mar. Freshw. Res. **35**: 213–259.
Entwisle T.J., Vis M.L., Chiasson W.B., Necchi Jr. O. & Sherwood A.R. (2009): Systematics of the Batrachospermales (Rhodophyta) – A synthesis. – J. Phycol. **45**: 704–715.
Feldmann J. (1953): L'évolution des organes femelles chez les Floridées. – Proc. First Int. Seaweed Symp. [Edinburgh], pp. 11–12.
Federer C.A. & Tanner C.A. (1966): Spectral disribution of light in the forest. – Ecology **47**: 555–560.
Fjerdingstad, E. (1964): Pollution of streams estimated by benthal phytomicroorganisms I. A Saprobic system based on communities of organisms and ecological factors. – Int. Rev. ges. Hydrobiol. **49**: 63–131.
Fjerdingstad, E. (1965): Taxonomy and saprobic valency of benthic phytomicroorganisms. – Int. Rev. ges. Hydrobiol. **50**: 475–604.
Flint L.H. (1947): Studies of freshwater red algae. – Amer. J. Bot. **34**: 125–13.
Flint L.H. (1948): Studies of freshwater red algae. – Amer. J. Bot. **35**: 428–433.
Flint L.H. (1949): Studies of freshwater red algae. – Amer. J. Bot. **36**: 549–552.
Flint L.H. (1950): Studies of freshwater red algae. – Amer. J. Bot. **37**:754–757.
Flint L.H. (1953): Two new species of *Batrachospermum*. – Proc. Louisiana Acad. Sci. **16**: 10–15.
Friedmann I. (1956): Beiträge zur Morphologie und Formwechsel der atmophytischen Bangioideae *Phragmonema sordidum*. – Österr. Bot. Zeitschr. **103**: 613–633.
Friedrich G. (1966): *Compsopogon hookeri* Montagne (Rhodophyceae, Bangioideae) neu für Deutschland. – Nova Hedwigia **12**: 399–403.
Fries E.M. (1825): Systema orbis vegetabilis. Pars I. Plantae homonemeae. – Lundae, pp. viii + 374.
Galdieri A. (1899): Su di unálga che cresce intorno alle fumarole della Solfatera. – Rendiconti Reale Accad. Sci. Fis. E Mat. [Napoli], ser. 3, **5**: 160–164.
Gantt E. (1977): Recent contributions in phycobiliproteins and phycobilisomes. – Photochem. Photobiol. **26**: 685–689.
Gantt E. (1990): Pigmentation and photoacclimation. – In: Cole K.M. & Sheath R.G. (eds.), Biology of the red algae: 203–219, Cambridge, University Press.
Gantt E. & Conti S.F. (1966): Phycobiliprotein localization in algae. – Brookhaven Symp. Biol. **19**: 393–405.
Gantt E., Lipschultz C.A., Grabowski J. & Zimmermann B.K. (1979): Phycobilisomes from blue-green and red algae. Isolation criteria and dissociation characteristics. – Plant Physiol. **63**: 615–620.

Gantt E., Scott, J. & Lipschultz C. (1986): Phycobiliprotein composition and chloroplast structure in the freshwater red alga *Compsopogon coeruleus* (Rhodophyta). – J. Phycol. **22**: 480–484.
Garbary D.J. (1987): The Acrochaetiaceae (Rhodophyta): an annotated bibliography. – Biblioth. Phycol. **77**: 1–267.
Garbary D.J. & Gabrielson P.W. (1990): Taxonomy and evolution. – In: Cole K.M. & Sheath R.G. (eds.), Biology of the red algae: 477–498, New York, Cambridge University Press.
Garbary D.J., Hansen G.I. & Scagel R.F. (1980): A revised classification of the Bangiophyceae (Rhodophyta). – Nova Hedwigia **33**: 145–146.
Geesink R. (1973): Experimental investigation on marine and freshwater *Bangia* (Rhodophyta) in the Netherlands. – J. Exp. Mar. Biol. Ecol. **11**: 239–247.
Geitler L. (1923): *Porphyridium aerugineum* n.sp. – Österr. Bot. Zeitschr. **72**: 84–85.
Geitler L. (1924): Über einige wenig bekannte Süsswasserorganismen mit roten oder blaugrünen Chromatophoren. – Rev. Algologique **4**: 357–375.
Geitler L. (1927): *Rhodospora sordida* nov. gen. et n. sp. eine neue *Bangiaceae* des Süsswassers. – Österr. Bot. Zeitschr. **76**: 25–28.
Geitler L. (1932): *Porphyridium sordidum* n. sp. eine neue Süsswasserbangiacee. – Arch. Protistenk. **76**: 595–604.
Geitler L. (1933): Diagnosen neuer Blaualgen von den Sunda-Inseln. – Arch. Hydrobiol. Suppl. **12**: 622–634.
Geitler L. (1944): Ein neues einheimisches *Batrachospermum* sowie Beobachtungen an anderen einheimischen *Batrachospermum*-Arten. – Österr. Bot. Zeitschr. **93**: 127–137.
Geitler L. (1970): Beiträge zur epiphytischen Algenflora des Neusiedlersees. – Österr. Bot. Zeitschr. **118**: 17–29.
Gobi Ch. (1879): Forschungen im Finnischen Meerbusen. – Arb. St. Petersb. Nat., Ges. **10**: 83–92.
Graham L.E., Graham J.M. & Wilcox L.W. (2009): Algae. 2nd Ed. Benjamin Cummings 616 pp. + appendix.
Greuter W., Burmitt R.K., Farr E., Killian N., Kirk P.M. & Silva P.C. (1993) (eds.): Names in Current Use for Extant Plant Genera-3. The International Association for Plant Taxonomy. – Koeltz Scientific Books, Königstein, pp. xxviii +1464.
Hambrook J.A. & Sheath R.G. (1987): Grazing of freshwater Rhodophyta. – J. Phycol. **23**: 656–662.
Hambrook J.A. & Sheath R.G. (1991): Reproductive ecology of the freshwater red alga *Batrachospermum boryanum* Sirodot in a temperate headwater stream. – Hydrobiologia **218**: 233–246.
Hamel G. (1925a): Floridées de France. III & IV. – Rev. Algol. **2**: 39–67, 280–309.
Hamel G. (1925b): Sur la synonymie des Chantransiées. – C. R. Congr. Soc. Savantes (Paris) pp. 239–241.
Hansgirg A. (1884): Bemerkungen zur Systematik einiger Süsswasseralgen. – Österr. Bot. Zeitschr. **9**: 18 pp.
Hansgirg A. (1885): Ein Beitrag zur Kenntnis von der Verbreitung der Chromatophoren und Zellkerne bei den Schizophyceen (Phytochromaceen). – Ber. Deutsch. Bot. Ges. **3**: 14–22.
Hanyuda T., Suzawa Y., Arai S., Ueda K. & Kumano S. (2004): Phylogeny and taxonomy of freshwater *Bangia* (Bangiales, Rhodophyta) in Japan. – J. Jpn. Bot. **79**: 262–268.

Harper J.T. & Saunders G.W. (2002): A re-classification of the Acrochaetiales based on molecular and morphological data, and establishment of the Colaconematales *ord. nov.* (Florideophyceae, Rhodophyta). – Europ. J. Phycol. **37**(3): 463–476.

Harvey W.H. (1841): A manual of the British marine algae. – London, pp. lvii + 229.

Harvey W.H. (1848): Phycologia Britannica 36: 211–216, London, Reeve & Benham.

Harvey W.H. (1858): Nereis borealis-americana. Part III. Chlorospermeae. – Smithsonian Contr. Knowl. **1** (2), pp. 140.

Hassall A.H. (1845): A history of the British freshwater algae. Vol. I & II. – London, pp. viii + 462, ciii plates + 24 pp. explan.

Helleburst J.A. (1970): Light-plants. – In: Kinne O. (ed.), Marine ecology, Vol. 1, Part 1: 125–148, London, Wiley Interscience.

Hirose H. (1958): Rearrangement of the systematic position of a thermal alga, *Cyanidium caldarium*. – Bot. Mag. Tokyo **71**: 347–352.

Hirose H. & Kumano S. (1966): Spectroscopic studies on the phycoerythrins from rhodophycean algae with special reference to their phylogenetical relations. – Bot. Mag. (Tokyo) **79**: 105–113.

Hirsch A. & Palmer C.M. (1958): Some algae from the Ohio River drainage basin. – Ohio J. Sci. **58**: 375–382.

Hoek van den C., Mann D.G. & Jahns H.M. (1995): Algae. An introduction to phycology. – Cambridge, University Press, pp. xiv + 623.

Holmes N.T. & Whitton B.A. (1981): Phytobenthos of the River Tees and its tributaries. – Freswater Biol. **11**: 139–163.

Honsell E., Kosovel V. & Talarico L. (1984): Phycobilin distribution in Rhodophyta: Studies and interpretation on the basis of their absorption spectra. – Bot. Mar. **27**: 1–16.

Hornemann J.W. (1818): Icones plantarum. Florae Danicae. Vol. 9, fasc. 27. – Copenhagen, pp. 1–11, plates MDLXI–MDCXX.

Hudson H. (1798): Flora Anglica. Editio III. – London, pp. XXXVIII + 690.

Israelson G. (1942): The freshwater Florideae of Sweden. – Symb. bot. upsaliensis **6**(1): 1–134.

Israelson G. (1949): Two freshwater Protoflorideae new to Sweden. – Bot. Notiser **1949**: 366–376.

Jao C.C. (1941): Studies on the freshwater algae of China. IV. Subaerial and aquatic algae from NanYoh, Huna. Part II. – Sinensia **11**: 241–361.

Jeffrey S.W. (1980): Algal pigments. – In: Falkowski P.G. (ed.), Primary productivity of the sea: 33–58, New York, Plenum Press.

John D.M., Whitton B.A. & Brook A.J. (2002) (eds.): The freshwater algal flora of the British Isles. An identification guide to freshwater and terrestrial algae. – Cambridge Univ. Press, Cambridge, pp. 702.

Kaczmarczyk D. & Sheath R.G. (1991): The effect of light regime on the photosynthetic apparatus of the freshwater red alga *Batrachospermum boryanum*. – Cryptogamie, Algol. **12**(4): 249–263.

Kaczmarczyk D. & Sheath R.G. (1992): Pigment content and carbon nitrogen ratios of freshwater red algae growing at different light levels. – Jpn. J. Phycol. (Sôrui) **40**: 279–282.

Kaczmarczyk D., Sheath R.G. & Cole K.C. (1992): Distribution and systematics of the freshwater genus *Tuomeya* (Rhodophyta, Batrachospermaceae). – J. Phycol. **28**: 850–855.

Kann E. (1978): Typification of Austrian streams concerning algae. – Verh. Intern. Verein. Limnol. **20**: 1523–1526.

Kapraun D.F. (2005): Nuclear DNA content estimates in multicellular green, red and brown algae: phylogenetics considerations. – Ann. Bot. **95**: 7–44.

Kapraun D.F., Braly K.S. & Freshwater D.W. (2007): Nuclear DNA content variation in the freshwater red algal orders Batrachospermales and Thoreales (Florideophyceae, Nemaliophycidae). – Phycologia **46**: 54–62.

Kerner A. (1882): Schedae ad floram exsiccatam Austro-Hungaricum a Museo botanico Universitatis Vindobonensis anno 1881 editam, fasc. II 8. – Bot. Centralb. **10**: 360–362.

Klas Z. (1936): Eine neue *Thorea* aus Jugoslavien, *Thorea brodensis* Klas sp. n. – Hedwigia **75**: 273–284.

Knappe J., Geissler U., Gutowski A. & Friedrich G. (1996): Rote Liste der limnischen Braunalgen (*Fucophyceae*) und Rotalgen (*Rhodophyceae*) Deutschlands. – Schr.-R. f. Vegetationskde. **28**: 609–623.

Korch J.E. & Sheath R.G. (1989): The phenology of *Audouinella violacea* (Acrochaetiaceae, Rhodophyta) in a Rhode Island stream (USA). – Phycologia **28**(2): 228–236.

Kostkevičienė J. & Laučiūtė R. (2005): The freshwater red algae species of *Batrachospermum* sect. *Batrachospermum* (*Batrachospermales*, *Rhodophyta*) new to Lithuania. – Bot. Lithuanica **11**(3): 151–159.

Kremer B. (1983): Untersuchungen zur Ökophysiologie einiger Süsswasserrotalgen. – Decheniana (Bonn) **136**: 31–42.

Kučera P. & Marvan P. (2004): Taxonomy and distribution of *Lemanea* and *Paralemanea* (*Lemaneaceae*, *Rhodophyta*) in the Czech Republic. – Preslia, Praha **76**: 163–174.

Kumano S. (1980): On the distribution of some freshwater red algae in Japan and southeast asia. – Proc. Workshop Prom. Limnol. Dev. Count. – **1**: 3–6.

Kumano S. (1993): Taxonomy of the family Batrachospermaceae (batrachospermales, Rhodophyta). – Jap. J. Phycol. **41**: 253–272.

Kumano S. (2002): Freshwater red algae of the World. – Dorset, Biopress Ltd., pp. 375.

Kumano S. (1993): Taxonomy of the family Batrachospermaceae (Batrachospermales, Rhodophyta). – Jap. J. Phycol. **41**: 253–272.

Kumano S. & Necchi Jr O. (1985): Studies on the freshwater Rhodophyta of Brazil. II. Two new species of *Batrachospermum* from states Amazonas and Minas Gerais. – Jap. J. Phycol. **33**: 181–189.

Kumano S. & Watanabe M. (1983): Two new varieties of *Batrachospermum* (Rhodophyta) from Mt. Albert Edward, Papua New Guinea. – Bull. Natl. Sci. Mus. (Tokyo), Ser. B, **9**: 85–94.

Kützing F.T. (1843): Phycologia generalis. – Leipzig, pp. xxxii + 458.

Kützing F.T. (1845): Phycologia germanica. – Nordhausen, pp. x + 340.

Kützing F.T. (1849): Species algarum. – Leipzig, pp. vi + 922.

Kützing F.T. (1857): Tabulae phycologicae; oder, Abbildungen der Tange. Vol. 7, Nordhausen, pp. ii + 40 + 100 plates.

Kwandrans J. & Eloranta, P. (1994): *Tuomeya americana* a freshwater red alga new to Europe. – Algol. Studies **74**: 27–33.

Kwandrans J. & Eloranta, P. (2002): Variation of the genus *Sirodotia* (Batrachospermales, Rhodophyta) in some Nordic waters. – Verh. Internat. Verein. Limnol. **28**: 1370–1372.

Kylin H. (1912): Studien über die schwedischen Arten der Gattungen *Batrachospermum* Roth und *Sirodotia* nov. gen. – Nova Acta Regiae Soc. Sci. Upsal., ser. 4, **3**(3): pp. 40.

Kylin H. (1930): Über die Entwicklungsgeschichte der Florideen. – Lunds Univ. Årsskr. N. F. Avd. 2, **26**(6): 1–104.

Kylin H. (1937): Über eine marine *Porphyridium*-Art. – Förh. Kongl. Fysiogr. Sällsk. Lund **7**(10): 119–123.

Kylin H. (1956): Die Gattungen der Rhodophyceen. – Gleerup Förlag, Lund, pp. 673.

Lamoroux J.V.F. (1813): Essai sur les genres de la famille des thalassiophytes non articulées. – Ann. Mus. Hist. Nat. (Paris) 20: 21–47, 115–139, 267–293.

Lee R.E. (1999): Phycology. – 3rd Ed. Cambridge, University Press, pp. 614.

Liebmann F. (1839): Om ett nytt genus *Erythrocladus* af algernas familie. – Nat. Tidskr. **2**: 169–175.

Liaaen-Jensen S. (1978): Marine carotenoids. – In: Scheuer P.J. (ed.), Marine natural products: Chemical and biological perspectives, Vol. 2: 1–73, New York, Academic Press.

Linnaeus C. (1753): Species plantarum. Vol. 2. – Stockholm, pp. 561–1200.

Lokhorst G.M. (1994): The red alga *Thorea ramosissima* Bory new for the Dutch flora. – Gorteria **20**: 5–14.

Lyngbye H.C. (1819): Tentamen hydrophytologiae danicae. – Hafnie, Copenhagen, pp. xxxii + 248.

Magne F. (1967a): Sur l'existence, chez les *Lemanea* (Rhodophycées, Némalionales), d'une type de cycle de développement encore inconnu chez les algues rouges. – C.R. Acad. Sci. [Paris] Sér. D, **264**: 2632–2633.

Magne F. (1967b): Sur le de déroulement et le lieu de la meiose chez les Lémanéacées (Rhodophycées, Némalionales). – C.R. Acad. Sci. [Paris] Sér. D, **265**: 670–673.

Manning W.M. & Strain H.H. (1943): Chlorophyll *d*, a green pigment in red algae. – J. Biol. Chem. **151**: 1–19.

Meneghini G. (1841): Lettera del Prof. Giuseppe Meneghini al Dott. Corinaldi. – Giorn. Tosc. Sci. Med. **1**: 186–189.

Merola A., Castaldo R., De Luca P., Gambardella R., Musacchio A. & Taddei R. (1982): Revision of *Cyanidium caldarium*. Three species of acidophilic algae. – Giorn. Bot. Ital. **115**: 189–195.

Möbius M. (1891, 1892): Beitrag zur Kenntnis der Gattung *Thorea*. – Ber. Deutsch. Bot. Ges. **9**: 333–343, **11**: 266–270.

Montagne C. (1846): Ordo I. Phyceae. – In: Durieu de Maisonneuve, M. C. (ed.), Exploration scientifique de l'Agérie pendant les années 1840, 1841, 1842. Science physique. Botanique. Cryptogamie. Paris, pp. 197.

Müller K.M., Cannone J.J., Gutell R.R. & Sheath R.G. (2001): A structural and phylogenetic analysis of the group IC1 introns in the order Bangiales (Rhodophyta). – Mol. Biol. Evol. **18**(9): 1654–1667.

Müller K.M., Cannone J.J. & Sheath R.G. (2005): A molecular phylogenetic analysis of the Bangiales (Rhodophyta) and description of a new genus and species, *Pseudobangia kaycoleia*. – Phycologia **44**: 146–155.

Müller K.M., Oliveira M.C., Sheath R.G. & Bhattacharya D. (2001): Ribosomal DNA phylogeny of the Bangiophycidae (Rhodophyta) and the origin of secondary plastids? Amer. J. Bot. **88**(8): 1390–1400.

Müller K.M., Sheath R.G., Vis M.L., Crease T.J. & Cole K.M. (1998): Biogeography and systematics of *Bangia* (Bangiales, Rhodophyta) based on the RuBisCo

spacer, *rbc*L gene and 18S rRNA gene sequences and morphometric analyses. 1. North America. – Phycologia **37**(3): 195–207.
Müller K.M., Sherwood A.R., Pueschel C.M., Gutell R.R. & Sheath R.G. (2002): A proposal for a new red algal order, the Thoreales. – J. Phycol. **38**: 807–820.
Müller K.M., Vis M.L., Chiasson W.B., Whittick A. & Sheath R.G. (1997): Phenology of a *Batrachospermum* population in a boreal pond and its implications for the systematics of section *Turfosa* (Batrachospermales, Rhodophyta). – Phycologia **36**(1): 68–75.
Nägeli C. (1849): Gattungen einzelliger Algen. – Neue Denkschr. Allgem. Schweiz. Ges. Naturwiss. **10**(7): 1–139.
Nardo J.D. (1834): De novo genre algarum cui nomen est Hildenbrandtia prototypus. – Isis (Oken) **6**: 675–676.
NCU-3 (1993): Names in current use for extant plant genera. – In: Greuter W., Brummit R.K., Farr E., Kilian N., Kirk P.M. & Silva P.C. (eds.): Regnum Vegetabile **129**: XXVI, pp. 1464.
Necchi O., Jr. (1986): Studies on the freshwater Rhodophyta of Brazil – 4. Four new species of *Batrachospermum* (section *Contorta*) from the southern state of São Paulo. – Revista Brasil. Biol. **46**: 517–525.
Necchi O., Jr. (1987): Sexual reproduction in *Thorea* Bory (Rhodophyta, Thoreaceae). – Jap. J. Phycol. **35**: 106–112.
Necchi O., Jr. (1988): Revisão de génerо *Batrachospermum* Roth (Rhodophyta, Batrachospermales) no Brasil. – Ph.D. Thesis. Universidade Estadual Paulista, Julio de Masquita Filho, Rio Caro.
Necchi O., Jr. (1990): Revision of the genus *Batrachospermum* Roth (Rhodophyta, Batrachospermales) in Brazil. – Biblioth. Phycol. **84**: 1–201.
Necchi O., Jr. (2005): Light-related photosynthetic characteristics of freshwater rhodophytes. – Aquatic Botany **82**(3): 193–209.
Necchi O., Jr. & Entwisle T.J. (1990): A reappraisal of generic and subgeneric classification in the Batrachospermaceae (Rhodophyta). – Phycologia **29**(4): 478–488.
Necchi O., Jr. & Zucchi M.R. (1995): Systematics and distribution of freshwater *Audouinella* (Acrochaetiaceae, Rhodophyta) in Brazil. – Eur. J. Phycol. **30**: 209–218.
Necchi O., Jr. & Zucchi M.R. (1997): Taxonomy and distribution of *Thorea* (Thoreaceae, Rhodophyta) in Brazil. – Arch. Hydrobiol. Suppl. **118** (Algol. Studies) **84**: 83–90.
Necchi O., Jr. & Zucchi M.R. (2001): Photosynthetic performance of freshwater Rhodophyta in response to temperature, irradiance, pH and diurnal rhythm. – Phycol. Res. **49**: 305–318.
Necchi O., Jr., Sheath R.G. & Cole K.M. (1993): Distribution and systematics of the freshwater genus *Sirodotia* (Batrachospermales, Rhodophyta) in North America. – J. Phycol. **29**: 236–243.
Necchi O., Jr., Sheath R.G. & Cole K.M. (1993a): Systematics of freshwater *Audouinella* (Acrochaetiaceae, Rhodophyta) in North America. 1. The reddish species. – Algol. Studies **70**: 11–28.
Necchi O., Jr., Sheath R.G. & Cole K.M. (1993b): Systematics of freshwater *Audouinella* (Acrochaetiaceae, Rhodophyta) in North America. 2. The bluish species. – Algol. Studies **71**: 13–21.
Nelson W.A. (2007): *Bangiadulcis* gen.nov.: a new genus for filamentous Bangiales (Rhodophyta). – Taxon **56**(3): 883–886.

Okamura K. (1915): Icones of Japanese algae. Vol. 3. – Tokyo, pp. 218 + plates ci–cl.
Ott F.D. & Seckbach J. (1994): *Cyanidium caldarium* (Rhodophyta) and related cells. – In: Seckbach J. (ed.), Evolutionary pathways and enigmatic algae: 145–152, Dordrecht, Kluwer Academic Publisher.
Ott F.D. & Seckbach J. (1994): New classification of the genus *Cyanidium* Geitler 1933. – In: Seckbach J. (ed.), Evolutionary pathways and enigmatic algae: 113–132, Dordrecht, Kluwer Academic Publisher.
Papenfuss G.F. (1958): Notes on algal nomenclature. IV. – Taxon **7**: 104–109.
Pascher A. (1925) (ed.): Die Süsswasser-Flora Deutschlands, Österreichs und der Schweiz. – Jena, Gustav Fischer Verlag.
Pascher A. & Petrova J. (1931): Über Porenapparate und Bewegung bei einer neuen Bangiale (*Chroothece mobilis*). – Arch. Protistenk. **74**: 490–522.
Pascher A. & Schiller J. (1925): Rhodophyta (Rhodophyceen). In: Pascher A. (ed.), Die Süsswasser-Flora Deutschlands, Österreichs und der Schweiz. Fasc. 11: 134–206, Jena, Gustav Fischer Verlag.
Pinto G. (2007): Cyanidiophyceae: Looking back – looking forward. – In: Seckbach J. (ed.), Algae and caynobacteria in extreme environments: 389–397, Springer.
Pritchard H.N. & Bradt P.T. (1984): Biology of nonvascular plants. – St. Louis, Times Mirror/Mosby, pp. 550.
Pueschel C.M. & Cole K.M. (1982): Rhodophycean pit plugs: an ultrastructural survey with taxonomic implications. – Amer. J. Bot. **69**: 703–720.
Rabenhorst L. (1859): Algen Sachsens. – Decades **83/84**, nos. 821–840.
Rabenhorst L. (1863): Kryptogamen-Flora von Sachsen. Vol. 1. – Leipzig, pp. xx + 653.
Raven J.A. & Beardall J. (1981): Carbon dioxide as the exogenous inorganic source for *Batrachospermum* and *Lemanea*. – Brit. Phycol. J. **16**: 165–175.
Raven J.A., Beardall J. & Griffiths H. (1982): Inorganic C-sources for *Lemanea*, *Cladophora* and *Ranunculus* in a fast-flowing stream: Measurements of gas exchange and of carbon isotope ratio and their ecological implications. – Oecologia (Berlin) **53**: 68–78.
Reis M.P. (1954): Contribuição para o conhecimento das espécies de *Batrachospermum* Roth da flora portuguesa. – Ciências (Madrid) **19**: 349–357.
Reis M.P. (1960): Revisão dos espécimes de Batrachospermum Roth e Sirodotia Kylin dos herbários dos institutos botanicos de Coimbra e Lisboa. – Bol. Soc. Brot., ser. 2, **34**: 37–55.
Reis M.P. (1962): Subsídios para o conhecimento das Rodoficeas de água doce de Portugal. III. – Mem. Soc. Brot. **15**: 57–71.
Reis M.P. (1965a): *Batrachospermum gulbenkianum*, sp. nov. – Anuário Soc. Brot. **31**: 31–44.
Reis M.P. (1965b): Subsídios para o conhecimento das Rodoficeas de água doce de Portugal. V. – Bol. Soc. Brot., ser. 2, **39**: 137–156.
Reis M.P. (1967): Duas espécies novas de *Batrachospermum* Roth: *B. azeredoi* e *B. ferreri*. – Bol. Soc. Brot., ser. 2, **41**: 167–189.
Reis M.P. (1970/71): Rhodophyceae novae. – Mem. Soc. Brot. **21**: 23–31.
Reis M.P. (1972a): Estudo comparativo de *Batrachospermum helminthosum* Bory, *B. coerulescens* Sirod. e *B. azeredoi* P. Reis e descrição de una variedade nova. – Bol. Soc. Brot., ser. 2, **46**: 191–217.
Reis M.P. (1972b): *Batrachospermum henriquesianum*, sp.nov. – Bol. Soc. Brot., ser. 2, **46**: 181–190.

Reis M.P. (1973): Subsídios para o conhecimento das Rodoficeas de água doce de Portugal. VIII. – Bol. Soc. Brot., ser. 2, **47**: 139–156.
Rider D.E. & Wagner R.H. (1972): The relationship of light, temperature, and current to seasonal distribution of *Batrachospermum* (Rhodophyta). – J. Phycol. **8**: 323–331.
Rieth A. (1979): Ein Standort der epiphytischen Süsswasser-Rotalge *Balbiana investiens* (Lenormand) Sidorot 1876 in Mitteleuropa. – Arch. Protistenk. **115**: 75–98.
Rintoul T., Sheath R.G. & Vis M.L. (1999): Systematics and biogeography of the Compsopogonales (Rhodophyta) with emphasis on freshwater genera in North America. – Phycologia **38**: 517–527.
Rosenvinge L.K. (1918): The marine algae of Denmark. Part. II. Rhodophyceae II (Cryptonemiales). – Kongel. Danske Vidensk. Selsk. Skr., Naturvidensk. og Math. Afd., **7**: 153–284.
Roth A.W. (1797): Catalecta botanica. Fasc. 1. – Leipzig, pp. viii + 244.
Roth A.W. (1800): Tentamen florae germanicae. Vol. 3, part 1. – Leipzig, pp. vii + 578.
Roth A.W. (1806): Catalecta botanica. Fasc. 2. – Leipzig.
Rott E., Hofmann G., Pall K., Pfister P. & Pipp E. (1997): Indikationslisten für Aufwuchsalgen. Teil 1: Saprobielle Indikation. – Bundesministerium für Land- und Forstwirtschaft, Wasserwirtschaftkataster, pp. 73.
Rott E., Pfister P., Van Dam H., Pall K., Pipp E., Binder N. & Ortler K. (1999): Indikationslisten für Aufwuchsalgen. Teil 2: Trophieindikation und autökologische Anmerkungen. – Bundesministerium für Land- und Forstwirtschaft, Wasserwirtschaftkataster, pp. 248.
Sabater S., Aboal M. & Cambra J. (1989): Nuevas observaciones de rodofíceas en aguas epicontinentales del NE y SE de España. – Limnetica **5**: 93–100.
Schagerl M. & Donabaum K. (2003): Patterns of major photosynthetic pigments in freshwater algae. 1. Cyanoprokaryota, Rhodophyta and Cryptophyta. – Int. J. Limnol. **39**: 35–47.
Schmitz F. (1896): Compsopogonaceae. – In: Engler A. & Prantl K. (eds.), Die natürlichen Pflanzenfamilien. I, Abt. 2: 318–320, Leipzig, Wilhelm Engelmann.
Schubert N., García-Mendoza E. & Pacheco-Ruiz I. (2006): Carotenoid composition of marine red algae. – J. Phycol. **42**: 1208–1216.
Setchell W.A. (1890): Concerning the structure and development of *Tuomeya fluviatilis* Harv. – Proc. Amer. Acad. Arts Sci., N.S. **17**: 53–68.
Sheath R.G. (1984): The biology of freshwater red algae. – Progress in Phycol. Res. **3**: 89–157.
Sheath R.G. (1986): Seasonality of phytoplankton in northern tundra ponds. – Hydrobiologia **138**: 75–83.
Sheath R.G. (2003): Red algae. – In: Wehr J.D. & Sheath R.G. (eds.), Freshwater algae of North America. Ecology and classification: 197–224, San Diego, Academic Press.
Sheath R.G. & Burkholder J.M. (1983): Morphometry of *Batrachospermum* populations intermediate between *B. boryanum* and *B. ectocarpum* (Rhodophyta). – J. Phycol. **19**: 324–33.
Sheath R.G. & Burkholder J.M. (1985): Characteristics of softwater streams in Rhode Island II. Composition and seasonal dynamics of macroalgal communities. – Hydrobiologia **128**: 109–118.
Sheath R.G., Burkholder J.M., Hambrook J.A., Hogeland A.M., Hoy E., Kane M.E., Morison M.O., Steinman A.D. & Van Alstyne K.L. (1986c): Characteris-

tics of softwater streams in Rhode Island. III. Distribution of macrophytic vegetation in a small basin. – Hydrobiologia **140**: 183–191.
Sheath R.G., Burkholder J.M., Morison M.O., Steinman A.D. & Van Alstyne K.L. (1986d): Effects of tree canopy removal by gypsy moth larvae on the macroalgae of a Rhode Island headwater stream. – J. Phycol. **22**: 567–570.
Sheath R.G. & Cole K.M. (1980): Distribution and salinity adaptations of *Bangia atropurpurea* (Rhodophyta), a putative migrant into the Laurentian Great Lakes. – J. Phycol. **16**: 412–420.
Sheath R.G. & Cole K.M. (1984): Systematics of *Bangia* (Rhodophyta) in North America.I. Biogeographic trends in morphology. – Phycologia **23**: 383–396.
Sheath R.G. & Cole K.M. (1990a): *Batrachospermum heterocorticum* sp. nov. and *Polysiphonia subtilissima* (Rhodophyta) from Florida spring-fed streams. – J. Phycol. **26**: 563–568.
Sheath R.G. & Cole K.M. (1990b): Differential Alcian blue staining in freshwater Rhodophyta. – Br. Phycol. J. **25**: 281–285.
Sheath R.G. & Cole K.M. (1992): Biogeography of stream macroalgae in North America. – J. Phycol. **28**: 448–460.
Sheath R.G. & Cole K.M. (1993): Distribution and systematics of *Batrachospermum* (Batrachospermales, Rhodophyta) in North America, 2. Chromosome numbers. – Phycologia **32**(4): 304–306.
Sheath R.G. & Hambrook J.A. (1988): Mechanical adaptations to flow in freshwater red algae. – J. Phycol. **24**: 107–111.
Sheath R.G. & Hambrook J.A. (1990): Freshwater ecology. – In: Cole K.M. & Sheath R.G. (eds.), Biology of the red algae: 423–453, New York, Cambridge University Press.
Sheath R.G., Hambrook J.A. & Nerone C.A. (1988): The benthic macro-algae of Georgian Bay, the North Channel and their drainage basin. – Hydrobiologia **163**: 141–148.
Sheath R.G., Hamilton P.B., Hambrook J.A. & Cole K.M. (1989): Stream macroalgae of the eastern boreal forest region of North America. – Can. J. Bot. **67**: 3553–3562.
Sheath R.G. & Hymes B.J. (1980): A preliminary investigation of the freshwater red algae in streams of southern Ontario, Canada. – Can. J. Botany **58**(11): 1295–1318.
Sheath R.G., Morison M.O., Cole K.M. & Van Alstyne K.L. (1986a): A new species of freshwater Rhodophyta, *Batrachospermum carpocontortum*. – Phycologia **25**(3): 321–330.
Sheath R.G., Morison M.O., Korch J.E., Kaczmarczyk D. & Cole K.M. (1986b): Distribution of stream macroalgae in south-central Alaska. – Hydrobiologia **135**: 259–269.
Sheath R.G. & Müller K.M. (1990): Distribution of stream macroalgae in four high arctic drainage basins. – Arctic **50**(4): 355–364.
Sheath R.G. & Müller K.M. (1999): Systematic status and phylogenetic relationships of the freshwater genus *Balbiania* (Rhodophyta). – J. Phycol. **35**: 855–864.
Sheath R.G., Müller K.M. & Sherwood A.R. (2000): A proposal of a new red algal order, the Thoreales. – J. Phycol. **36** (Suppl. 3): 62.
Sheath R.G., Müller K.M., Vis M.L. & Entwisle T.J. (1996a): A re-examination of the morphology, ultrastructure and classification of genera in the Lemaneaceae (Batrachospermales, Rhodophyta). – Phycol. Res. **44**: 233–246.
Sheath R.G. & Sherwood A.R. (2002): Phylum Rhodophyta (Red algae). – In: John D.M., Whitton B.A. & Brook A.J. (eds.), The freshwater algal flora of the British

Isles. An identification guide to freshwater and terrestrial algae: 124–143, Cambridge, University Press.
Sheath R.G. & Vis M.L. (1995): Distribution and systematics of *Batrachospermum* (Batrachospermales, Rhodophyta) in North America. 7. Section *Hybrida*. – Phycologia **34**(5): 431–438.
Sheath R.G., Vis M.L., Hambrook J. A. & Cole K.M. (1996b): Tundra stream macroalgae of North America: composition, distribution and physiological adaptations. – Hydrobiologia **335**: 67–82.
Sheath R.G., Vis M.L. & Cole K.M. (1992): Distribution and systematics of *Batrachospermum* (Batrachospermales, Rhodophyta) in North America. 1. Section *Contorta*. – J. Phycol. **28**: 237–246.
Sheath R.G., Vis M.L. & Cole K.M. (1993a): Distribution and systematics of *Batrachospermum* (Batrachospermales, Rhodophyta) in North America. 3. Section *Setacea*. – J. Phycol. **29**: 719–725.
Sheath R.G., Vis M.L. & Cole K.M. (1993b): Distribution and systematics of red algal family Thoreaceae in North America. – Eur. J. Phycol. **28**: 231–241.
Sheath R.G., Vis M.L. & Cole K.M. (1993c): Distribution and systematics of freshwater Ceramiales (Rhodophyta) in North America. – J. Phycol. **29**: 108–117.
Sheath R.G., Vis M.L. & Cole K.M. (1994a): Distribution and systematics of *Batrachospermum* (Batrachospermales, Rhodophyta) in North America. 4. Section *Virescentia*. – J. Phycol. **30**: 108–117.
Sheath R.G., Vis M.L. & Cole K.M. (1994b): Distribution and systematics of *Batrachospermum* (Batrachospermales, Rhodophyta) in North America. 5. Section *Aristata*. – Phycologia **33**(6): 404–414.
Sheath R.G., Vis M.L. & Cole K.M. (1994c): Distribution and systematics of *Batrachospermum* (Batrachospermales, Rhodophyta) in North America. 6. Section *Turfosa*. – J. Phycol. **30**: 872–884.
Sheath R.G., Whittick A. & Cole K.M. (1994d): *Rhododraparnaldia oregonica*, a new freshwater red algal genus and species intermediate between the Acrochaetiales and the Batrachospermales. – Phycologia **33**(1): 1–7.
Sherwood A.R. & Sheath R.G. (1999): Biogeography and systematics of *Hildenbrandia* (Rhodophyta, Hildenbrandiales) in North America: inference from morphometrics and *rbc*L and 18S rRNA gene sequence analyses. – Eur. J. Phycol. **34**: 523–532.
Sherwood A.R. & Sheath R.G. (2000): Biogeography and systematics of *Hildenbrandia* (Rhodophyta, Hildenbrandiales) in Europe: inference from morphometrics and *rbc*L and 18S rRNA gene sequence analyses. – Eur. J. Phycol. **35**: 143–152.
Sherwood A.R. & Sheath R.G. (2003): Systematics of the Hildenbrandiales (Rhodophyta): gene sequence and morphometric analyses of global collections. – J. Phycol. **39**: 409–422.
Sherwood A.R., Shea T.B. & Sheath R.G. (2002): European freshwater *Hildenbrandia* (Hildenbrandiales, Rhodophyta) have not been derived through multiple invasions from marine habitats – Phycologia **41**: 87–95.
Sherwood A.R., Vis M.L., Entwisle T.J., Necchi Jr, O. & Presting G.G. (2008): Contrasting intra versus interspecies DNA sequence variation for representatives of the Batrachospermales (Rhodophyta): Insights from a DNA barcoding approach. – **56**(4): 269–279.
Silva P.C. (1959): Remarks on algal nomenclature. II. – Taxon **8**: 60–64.

Sirodot M.S. (1872): Étude anatomique, organogénique et physiologique sur les algues d'eau douce de la famille des Lémanéacées. – Ann. Sci. Natur., Ser. 5, Botanique **16**: 1–95 + 8 plates.
Sirodot M.S. (1873a): Nouvelle classification des algues d'eau douce du genre *Batrachospermum*; développement; generations alternantes. – Compt. Rend. Acad. Sci. [Paris] **76**: 1216–1220.
Sirodot M.S. (1873b): Développement des algues d'eau douce du genre *Batrachospermum*; generations alternantes. – Compt. Rend. Acad. Sci. [Paris] **76**: 1335–1339.
Sirodot M.S. (1874): Observations sur les phénomènes essentiels de la fécondation chez les algues d'eau douce du genre *Batrachospermum*. – Compt. Rend. Acad. Sci. [Paris] **79**: 1366–1369.
Sirodot M.S. (1875): Observations sur le développement des algues d'eau douce composant le genre *Batrachospermum*. – Bull. Bot. Soc. France **22**: 128–145.
Sirodot M.S. (1876): Le *Balbiania investiens*. Étude organogénique et physiologique. – Ann. Sci. Nat. Bot., ser. 6, **3**: 146–174.
Sirodot M.S. (1884): Les Batrachospermes: organisation, fonctions, développement, classification. – Paris, pp. 299 + 50 plates.
Skuja H. (1925): Bemerkungen über die Süsswasserarten der Gattung *Lithoderma* Aresch. in Lettland. – Hedwigia **65**: 331–340.
Skuja H. (1926a): Eine neue Süsswasserbangiaceae *Kyliniella* nov. gen. nov. sp. – Acta Horti Bot. Univ. Latv. **1**: 1–6.
Skuja H. (1926b): Zur Verbreitung und Ökologie von *Hildenbrandia rivularis* (Liebm.) Bréb. in Lettland. – Acta Univ. Latv. **14**: 659–672.
Skuja H. (1928): Vorarbeiten zu einer Algenflora von Lettland. IV. – Acta Horti Bot. Univ. Latv. **3**: 103–218.
Skuja H. (1931): Untersuchungen über die Rhodophyceen des Süsswassers. [I, II]. – Arch. Protistenk. **74**: 297–308.
Skuja H. (1933): Die Batrachospermaceen und Lemaneaceen Finnlands. – Memor. Soc. F. Fl. Fenn. **9**: 139–141.
Skuja H. (1934): Untersuchungen über die Rhodophyceen des Süsswassers. [IV– VI]. – Beih. Bot. Centralbl., Abt. B, **52**: 173–192.
Skuja H. (1938a): Comments on fresh-water Rhodophyceae. – Bot. Rev. **4**: 665–676.
Skuja H. (1938b): Die Süsswaserrhodophyceen der Deutschen Limnologischen Sunda-Expedition. – Arch. Hydrobiol., Suppl. **15**: 603–637.
Skuja H. (1939): Versuch einer systematischen Eintelung der Bangioideen oder Protoflorideen. – Acta Horti Bot. Univ. Latv. **11/12**: 23–38.
Skuja H. (1964): Grundzüge der Algenflora und Algenvegetation der Fjeldgegenden um Abisko in Schwedisch-Lappland. – Nova Acta R. Soc. Scient. upsal. Ser. 4, **18**(3): 1–465.
Skuja H. (1969): Eigentümliche morphologische Anpassung eines *Batrachospermum* gegen mechanische Schädigung in fliessenden Wasser. – Österr. Bot. Z. **116**: 55–64.
Sládeček V. (1973): System of water quality from the biological point of view. – Arch. Hydrobiol., Beih. **7**: 1–218.
Smith G.M. (1933): The fresh-water algae of the United States. – New York, McGraw-Hill, pp. xi + 716.
Smith J.H.C. & Benitez A. (1955): Chlorophylls: Analysis in plant materials. – In: Paech K. & Tracey M.V. (eds.), Modern methods of plant analysis, Vol. 4: 142–190, Berlin, Springer Verlag.

Starmach K. (1928): Beitrag zur Kenntnis der Süsswasserflorideen von Polen. – Acta Soc. Bot. Pol. **5**: 367–389.
Starmach K. (1977): Flora Słodkowodna Polski. Tom 14. Phaeophyta – Brunatnice, Rhodophyta – Krasnorosty. – Warszawa, PWN, pp. 445.
Starmach K. (1984a): Red algae in the Kryniczanka stream. – Fragm. Flor. Geobot. **28**: 257–293.
Starmach K. (1984b): New species to Poland and new localities of some species of *Batrachospermum* Roth (Rhodophyceae). – Fragm. Flor. Geobot. **28**: 303–309.
Starmach K. (1985): *Chantransia hermannii* (Roth) Desvaux and the systematic position of the genera *Chantransia*, *Pseudochantransia* and *Audouinella*. – Acta Soc. Bot. Polon. **54**: 273–284.
Stewart S.A. & Vis M.L. (2007): Investigation of two species complexes in *Batrachospermum* section *Batrachospermum* (Batrachospermales, Rhodophyta). – Phycologia **46**(4): 380–385.
Strain H.H. (1958): Chloroplast pigments and chromatographic analysis. – 32nd Annual Priestly Lecture, University Park, PA, Penn. State University Press.
Strain H.H. (1966): Fat-soluble chloroplast pigments: Their identification and distribution in various Australian plants. – In: Goodwin T.W. (ed.), Biochemistry of chloroplasts **1**: 387–406, London, Academic Press.
Swale E.M.F. & Belcher J.H. (1963): Morphological observations on wild and cultured material of *Rhodocorton investiens* (Lenormand) nov. comb. (*Balbiania investiens* (Lenorm.) Sirodot). – Ann. Bot. (London), ser. 2, **27**: 281–290.
Taylor W.R. (1957): Marine algae of the northeastern coast of North America. 2nd ed., Ann Arbor, University of Michigan Press, pp. ix + 509.
Thirb H.H. & Benson-Evans K. (1985): The effect of water temperature, current velocity and suspended solids on the distribution, growth and seasonality of *Lemanea fluviatilis* (C.Ag.), Rhodophyta, in the River Usk and other South Wales rivers. – Hydrobiologia **127**: 63–78.
Thore J. (1799): Sur la *Conferva hispida*. – Mag. Encycl., An 5, **6**: 398–403.
Tilden J.E. (1898): Observations on some West American thermal algae. – Bot. Gaz. **25**: 89–105.
Tomas X., Lopez P., Margalef-Mir R. & Comin F.A. (1980): Distribution and ecologyof *Compsopogon coeruleus* (Balbis) Montagne (Rhodophyta, Bangiophycidae) in Eastern Spain. – Cryptog. Algol. **1**: 179–186.
Vis M.L., Carlson T.A. & Sheath R.G. (1991): Phenology of *Lemanea fucina* (Rhodophyta) in a Rhode Island river, USA. – Hydrobiologia **22**: 141–146.
Vis M.L. & Entwisle T.J. (2000): Insights into the phylogeny of the Batrachospermales (Rhodophyta) from *rbc*L sequence data of Australian taxa. – J. Phycol. **36**: 1175–1182.
Vis M.L., Harper J.T. & Saunders G.W. (2007): Does the combination of large subunit rDNA and *rbc*L gene sequence data provide more resolution among taxa of the Batrachospermales (Rhodophyta)? – Phycol. Res. **55**: 103–112.
Vis M.L., Hodge J.C. & Necchi O. Jr. (2008): Phylogeography of *Batrachospermum macrosporum* (Batrachospermales, Rhodophyta) from North and South America. – J. Phycol. **44**: 882–888.
Vis M.L., Miller E.J. & Hall M.M. (2001): Biogeographical analyses of *Batrachospermum helminthosum* (Batrachospermales, Rhodophyta) in North America using molecular and morphological data. – Phycologia **40**: 2–9.
Vis M.L., Saunders G.W., Sheath R.G., Dunse K. & Entwisle T.J. (1998): Phylogeny of the Batrachospermales (Rhodophyta) inferred from rbcL and 18S ribosomal DNA gene sequences. – J. Phycol. **34**: 341–350.

Vis M.L. & Sheath R.G. (1992): Systematics of the freshwater red algal family Lemaneaceae in North America. – Phycologia **31**(2): 164–179.
Vis M.L. & Sheath R.G. (1993): Distribution and systematics of *Chroodactylon* and *Kyliniella* (Porphyridiales, Rhodophyta) from North American streams. – Jpn. J. Phycol. (Sôrui) **41**: 237–241.
Vis M.L. & Sheath R.G. (1996): Distribution and systematics of *Batrachospermum* (Batrachospermales, Rhodophyta) in North America. 9. Section *Batrachospermum*: description of five new species. – Phycologia **35**(2): 124–134.
Vis M.L. & Sheath R.G. (1997): Biogeograhy of *Batrachospermum gelatinosum* (Batrachospermales, Rhodophyta) in North America based on molecular and morphological data. – J. Phycol. **33**: 520–526.
Vis M.L. & Sheath R.G. (1999): A molecular investigation of the systematic relationships of *Sirodotia* species (Batrachospermales, Rhodophyta) in North America. – Phycologia **38**(4): 261–266.
Vis M.L., Sheath R.G., & Cole K.M. (1992): Systematics of the freshwater red algal family Compsopogonaceae in North America. – Phycologia **31**(6): 564–575.
Vis M.L., Sheath R.G., & Cole K.M. (1996): Distribution and systematics of *Batrachospermum* (Batrachospermales, Rhodophyta) in North America. 8a. Section *Batrachospermum*: *Batrachospermum gelatinosum*. – Eur. J. Phycol. **31**: 31–40.
Vis M.L., Sheath R.G. & Entwisle T.J. (1995): Morphometric analysis of *Batrachospermum* Section *Batrachospermum* (Batrachospermales, Rhodophyta) type specimens. – Eur. J. Phycol. **30**: 35–55.
Vis M.L., Hambrook J. A. & Cole K.M. (1994): Stream macroalgae of the Hawaiian Islands: A preliminary study. – Pacific Science **48**(2): 175–187.
Walter H. (1961): Einführung in die Botanik. Bd. II. Grundlagen des Pflanzensystems. – 3rd ed., Stuttgart, Eugen Ulmer, pp. 280.
Weber-van Bosse A. (1921): Liste des algues du Siboga. II. Rhodophyceae. Première partie. Protoflorideae, Nemalionales, Cryptonemaliales. – Siboga-Exped. Monogr. **59**: 187–310.
Webster R.N. (1958): The life history of the freshwater alga *Tuomeya fluviatilis* Harvey. – Butler Univ. Bot. Stud. **13**: 141–159.
West W. & West G.S. (1897): Welwitsch's African freshwater algae. – J. Bot. **35**: 1–7, 33–42, 77–89, 113–122, 172–183, 235–243, 264–272, 287–304.
Whitton B.A. (1975): Algae. – In: Whitton B.A. (ed.), River ecology: 81–105, Berkeley & Los Angeles, California University Press.
Wołowski K., Kowalska J. & Hindák F. (2007): *Chroodactylon ornatum* (Rhodophyta, Porphyridiales) occurring in Poland and Slovakia. – Biologia, Bratislava **62**(6): 646–649.
Wołozyńska J. (1935): *Bangia atropurpurea* (Roth) Ag. w Polsce – *Bangia atropurpurea* (Roth) Ag. in Polen. – Arch. Hydrobiol. Ryb. **9**: 139–162.
Woolcott G.W. & King R.J. (1998): Porphyra and Bangia (Bangiaceae, Rhodophyta) in warm temperate waters of eastern Australia: Morphology and molecular analyses. – Phycol. Res. **46**: 111–123.
Xie S. & Shi Z. (2004): Taxonomy of algal genus *Sirodotia* Kylin (Batrachospermaceae, Rhodophyta) in China. – J. Trop. Subtrop. Bot. **12**(1): 1–6.
Yoon H.S., Müller K.M., Sheath R.G., Ott F.D. & Bhattacharya D. (2006): Defining the major lineages of red algae (Rhodophyta). – J. Phycol. **42**: 482–492.
Žáková Z., Pum M., Sedláček & Hindák F. (2010): Výskyt ruduchy *Compsopogon* sp. (Rhodophyta) v Pulkavě, rakouském řeky Dyje. [Occurrence of a red alga

Compsopogon sp. (Rhodophyta) in the Pulkau/Pulkava River (Austria), a tributary of the Thaya/Dyje River]. – Limnol. Sprav. **4**(1): 10–14.

Zopf W. (1882): Zur Morphologie der Spaltpflanzen (Spaltpilze und Spaltalgen). – Veit & Comp., Leipzig, pp. vi + 74.

Zucchi M.R. & Necchi O., Jr. (2003): Blue-greenish acrochaetioid algae in freshwater habitats are "Chantransia" stages of Batrachospermales "sensu lato" (Rhodophyta). – Cryptogamie, Algol. **24**(2): 117–131.

Part II Phaeophyceae

Elsalore Kusel-Fetzmann

Introduction

Brown algae are characterized by their brownish colour caused by the accessory pigment fucoxanthin. This carotenoid and other pigments (e.g., violaxanthin, neoxanthin, diadinoxanthin, β-carotene) are located within the chloroplasts, masking the greenish colour of chlorophylls a, c_1 and c_2. The **chloroplasts** (formerly termed phaeoplasts based on their brown colour) are either solitary or several can be present per cell; they can be discoid or ribbon like. One or several pyrenoids – small rounded bodies attached to the inner side of the chloroplasts – may occur. The ultrastructure of the chloroplasts is comparable to that of other Heterokontophyta: three stacked thylakoids form a so-called lamella; the stacks are enclosed by a girdle lamella. Chloroplasts are surrounded by two plastid membranes and two periplastidial folds of endoplasmatic reticulum, which is connected with the nuclear envelope.
The photosynthetic storage product is the hydro-soluble β-1,3 glucan laminarin (Chrysolaminarin), which is accumulated dissolved in cytoplasmatic vacuoles; oil droplets and the sugar alcohol mannitol may also occur (Van den Hoek et al. 1995). Other small light-refractive vesicular bodies (physodes) contain fucosan, a phlorotannin (phloroglucinol polymers). Additionally, several marine Phaeophyceae store iodine in such high amounts that it was formerly extracted for industrial iodine production (Hoppe 1962).
Cell walls of brown algae are composed of a network of fibrillar cellulose embedded in an amorphous matrix of colloid alginates (salts of alginic acid). In some marine species, alginates are produced in such quantities that they are commercially harvested for food and pharmaceutical industries because of their colloidal nature (Schmid & Hoppe 1962, Lüning 1985).
Cell division (mitosis) is semi-closed (in the late anaphase the nuclear membrane disappears). In prophase, the paired centrioles, lying at the poles outside of the nucleus membrane, are the centres of radiating microtubules. In metaphase the spindle apparatus is formed through holes of the nucleus envelope ending in the centrioles. At metaphase the chromosomes lie in a plane equatorial to the spindle. After telophase, in cytokinesis, the cells are separated by inward furrowing of the plasmalemma. The cross wall is deposited in the furrow. Numerous plasmodesmata penetrate the wall for connection and transport between the cells (Van Den Hoek et al. 1995, Lee 1999).
Two types of **thallus morphology** can be distinguished: (1) parenchymatous macrothalli with distinct meristematic growth zones. Marine seaweeds from bushy, foliose to stem-like kelps, including the largest known seaweeds, are placed in this group. (2) Thin uniseriate filaments, loosely prostrate creeping (or compact filaments as pseudoparenchymatous crustose patches) and erect filaments as tufts or bundles. All freshwater taxa belong to the latter type and reach in height only several millimetres to a few centimetres.
The filaments are uniseriate *(Ectocarpus, Bodanella, Heribaudiella, Pleurocladia, Porterinema)* to multiseriate *(Sphacelaria)*. Creeping, prostrate filaments form discs or crusts by apical cell division, from which branched or unbranched erect

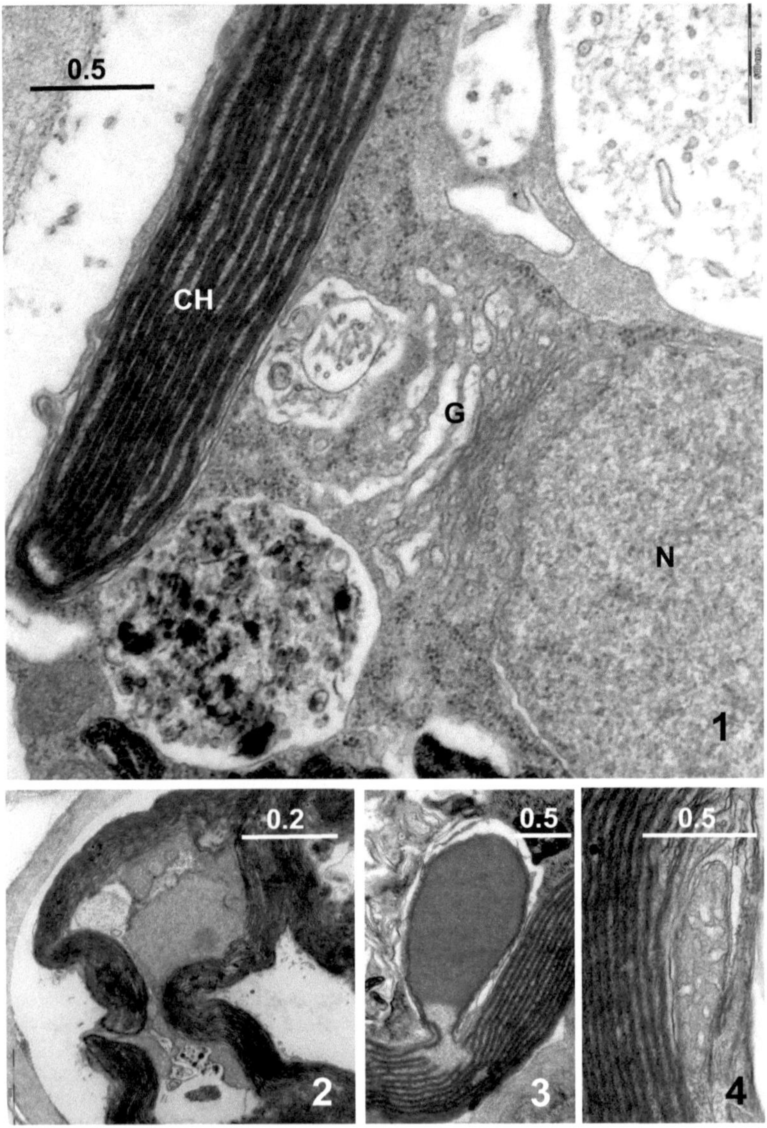

Fig. 8. Ultrastructure of *Porterinema fluviatile*. **1**: Chloroplast (CH) with thylakoids in triplets, surrounded by a girdle lamella. Nucleus (N) with attached golgy body (G). **2**: Cell dividing into a plurilocular sporangium. **3**: Pear shaped pyrenoid attached to the inner side of the chloroplast. **4**: Mitochondrion located close to the chloroplast.

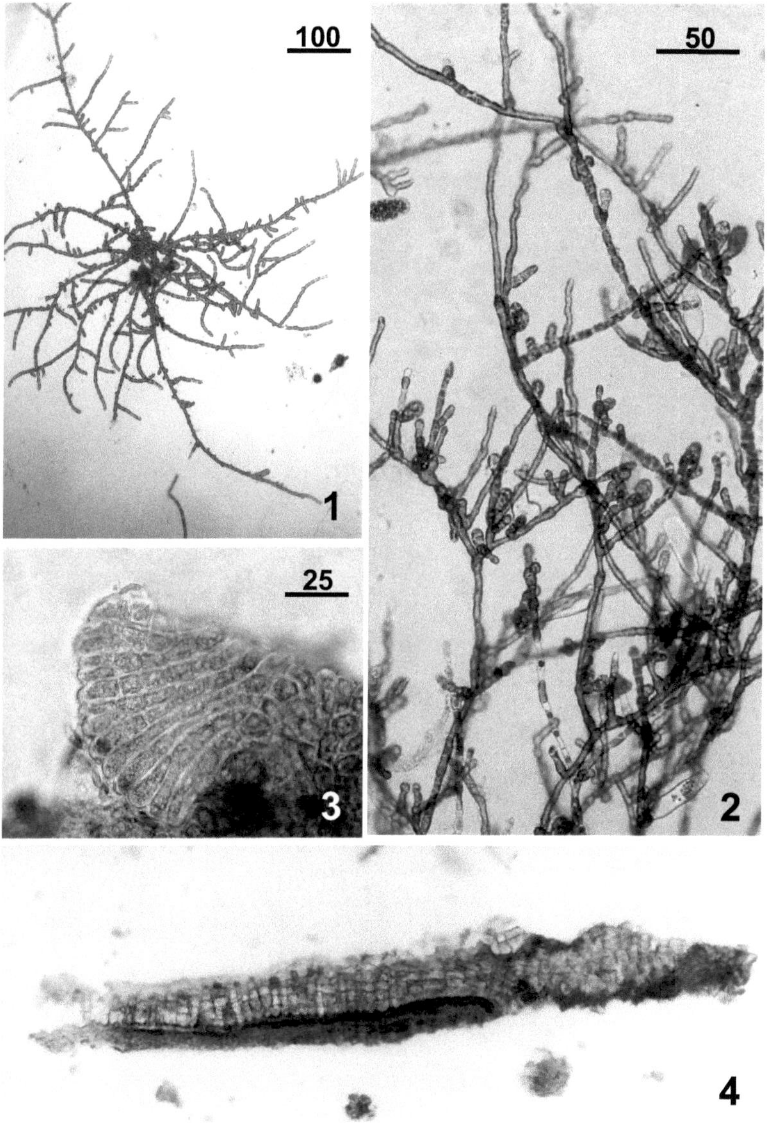

Fig. 9. Growth shape of freshwater species. **1**: Prostrate creeping filaments, *Porterinema fluviatile*. **2**: Upright bunch, one sided branches, unilocular sporangium, *Pleurocladia lacustris*. **3, 4**: *Heribaudiella fluviatilis*. **3**: young crust with growing edge, one layered. **4**: cross section of an older crust, showing the densely packed upright filaments.

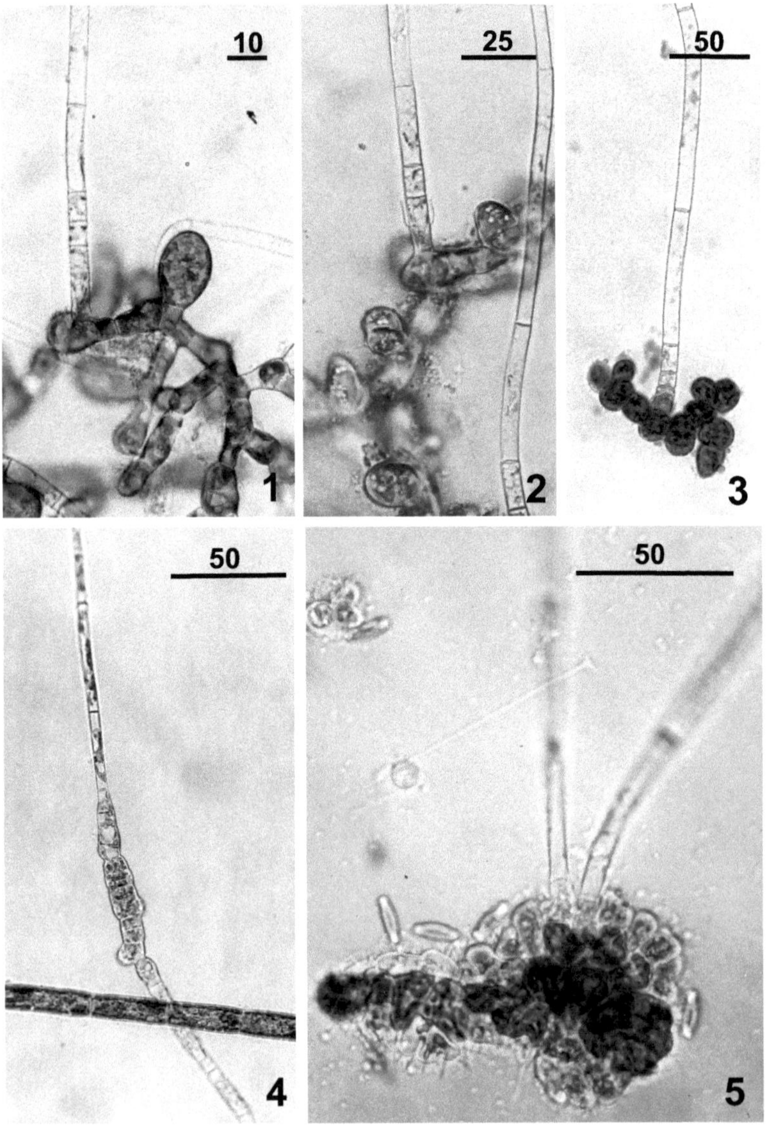

Fig. 10. Phaeophycean hairs with a basal meristem sheated by a husk. **1**: *Pleurocladia lacustris*. **2**: *Porterinema fluviatile*. **3**: *Bodanella lauterborni*. **4**: *Ectocarpus* with "false hair", which is a thin filament without basal meristem produced as an extension of a filament or sporangium. **5**: *Heribaudiella fluviatilis*, young crust with two hairs.

filaments arise, growing with intercalary meristematic zones. Thin pale uniseriate and unbranched hairs may develop with basal meristems.

Classification and systematics

Classification of the Phaeophyceae follows morphological characteristics, thallus organisation, life cycles, modes of growth and types of sexual reproduction. Accordingly, about 39 orders in marine Phaeophyceae were established in the past (Silva & De Reviers 2000). Today, 13–17 orders are accepted, depending on the researcher´s opinion (for an overview see Draisma et al. 2001). The freshwater brown algae show only few differential characters and life cycles are incompletely known.

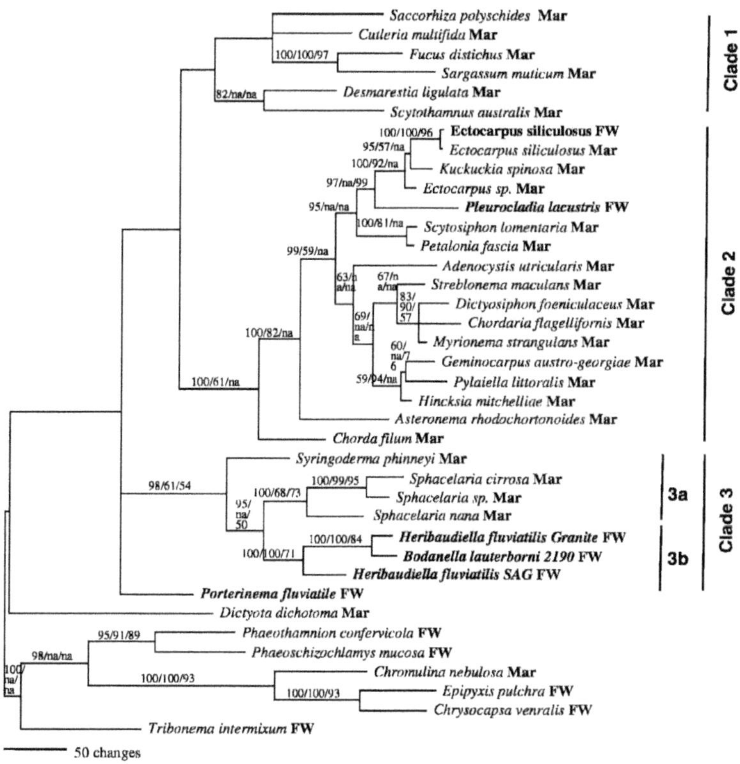

Fig. 11. Bayesian consensus tree for the alignment of rbcL sequences from freshwater (FW) and marine (Mar) Phaeophyceae. Values at the branches indicate Bayesian posterior probabilities, maximum parsimony bootstrap, and neighbour-joining bootstrap support, respectively (na = <40% support by that method; from McCauly & Wehr 2007).

With the exception of *Sphacelaria*, all freshwater Phaeophyceae have been classified as Ectocarpales (Bourrelly 1981, Pankow 1971, Starmach 1977, Pueschel & Stein 1983).

DNA sequence data have yielded fundamental new insights into the phylogenetic relationships within marine Phaeophyceae (De Reviers et al. 2007). The first molecular study of freshwater Phaeophyceae was done by Mc Cauley & Wehr (2007) and included *Bodanella lauterborni, Ectocarpus siliculosus, Heribaudiella fluviatilis, Pleurocladia lacustris* and *Porterinema fluviatile*, all of them originally considered as Ectocarpales.

These were compared with an existing nucleotide database from marine brown algae. Sequencing of the large subunit of the ribulose-bisphosphate carboxylase (rbcL) gene analysis resulted in three clades. Clade 1 comprises only marine species. Clade 2 comprises all members of Ectocarpales, including *Pleurocladia* and the freshwater *Ectocarpus siliculosus* from Australia. Clade 3 was divided in two subclades: 3a included marine *Sphacelaria* and 3b included the freshwater species *Heribaudiella* and *Bodanella*, which could be related to other crust-forming taxa, although for these taxa rbcL sequences were currently not available. *Porterinema* did not group closely with any of the species tested, so its position remains unclear. Therefore, *Heribaudiella*, *Bodanella* and *Porterinema* should remain unclassified as "incertae sedis".

In the following description of the freshwater Phaeophyceae, the positions within orders are neglected because further studies have to clarify the molecular phylogenies of Phaeophyceae in general.

Reproduction

Flagellates (zooids, zoospores, gametes) are the only unicellular stages in the life cycle of brown algae, and these structures are always reproductive. The flagellates show two laterally inserted flagellas, whereby the one bearing two rows of mastigonemes (= pleuronematic) is directed forward. The second, shorter flagellum without hairs (= acronematic) is directed to the posterior pole. In many cases, flagellates possess a photoreceptor swelling on the acronematic flagellum. Within the chloroplast, opposite the photoreceptor, an eyespot (stigma) is located, which acts as shade dispenser.

Sexual reproduction takes place via diplo-haplonts or diplonts. The latter are found in the order Fucales: here, the haploid generation is reduced to the egg and sperm cells in special cavities (conceptacles) at the surface of the diploid macrothallus. After release and fusion of the gametes, the zygote grows to a new diploid thallus. The diplo-haplont generation cycle includes a change from the diploid sporophyte to the haploid gametophyte; both isomorphic and heteromorphic cycles are found. Many haploid meiospores are produced in unilocular sporangia of diploid sporophytes. These meiospores swarm, settle and eventually grow to become haploid gametophytes. In plurilocular gametangia, one gamete per chamber is produced. Male and female gametes may be of the same size (isogamous), although sometimes the female gametes are larger than male gametes (heterogamous), or females are immobile without flagellas (oogamous taxa). In many cases, unilocular sporangia and plurilocular structures also produce asexual spores (Fritsch 1965).

For freshwater taxa, parts of the sexual reproduction cycle are still unknown. An asexual propagation without flagellated spores is found in *Sphacelaria*: multicellular triradiate propagules germinate to a new thallus after dropping off from the thallus.

Distribution and Ecology

We know little about the **geographic distribution** of freshwater Phaeophyceae. The taxa are usually small and inconspicuous and can be found only by thorough search and the use of microscopes. Some species were observed only sporadically, making any new findings worthy of publication.

Pleurocladia lacustris was first recognized by A. Braun (1854) as an epiphyte in Lake Tegeler See near Berlin (Germany). In the meantime, *Pleurocladia* is well known in Europe, and from Ukraine till Caspian Sea in Asia (Borisova 2006), further USA and Canada.

Heribaudiella fluviatilis is reported from many countries in Europe (Austria, France, Germany, Poland, Sweden, Ukraine, United Kingdom), China, Japan, Siberia, Canada, USA, but not from Africa, South America, Australia or New Zealand (see also the comprehensive list in Wehr & Stein 1985).

Other species show a disjunct distribution. *Sphacelaria fluviatilis*, for example, was found only in a Chinese river (Jao 1943) and in Gull Lake in Michigan, USA (Thompson 1975). *Sphacelaria lacustris* has been described only once from 5–15 m depth in western Lake Michigan (Schloesser & Blum 1980).

Ecology: Generally, freshwater brown algae are benthic forms growing epilithically, epiphytically or endophytically in running waters or lakes. The most widespread taxon is *Heribaudiella*, which inhabits gravel, boulders and rocks in fast-flowing, clear-water rivers, from a few centimetres down to two metres depth (e.g., in the Rhine River, Backhaus 2006), or on steep banks in lakes down to 30 m depth. In many cases, *Heribaudiella* and the rhodophyte *Hildenbrandia* occur together ("*Heribaudiella – Hildenbrandia* association" Fritsch 1929, Geitler 1932, Symoens 1950), but sometimes *Heribaudiella* is found without *Hildenbrandia* (Budde 1927). *Pleurocladia lacustris* – another widely distributed species – settles on stones, on *Heribaudiella* crusts in small streams, and in lakes as an epiphyte on *Phragmites* stalks (Waern 1952, Kirkby et al. 1972, Kann 1993, Szymanska & Zakrys 1990). *Pleurocladia* spores also germinated on artificial substrates (Polyethylene, Polyester, Polypropylene, PVC), and were documented for Lake Traunsee for the first time using that approach (Kann & Tschamler 1976).

In general, the **light requirements** of phaeophyta are described as low. In clear running waters, the light supply can be reduced by canopy cover, structures such as bridges, or by the river morphology itself (e.g., by small canyons). *Heribaudiella* prefers such shaded sites. *Heribaudiella* and *Pleurocladia,* however, were also reported from bright, sunny locations (Wehr & Stein 1985, Wehr & Perrone 2003). With increasing water depth, both light quantity and light quality change (Oberdorfer 1928). Many algal species are unable to cope with greenish light conditions, which commonly occur in deep water. This is the niche for taxa like *Bodanella lauterborni* and *Sphacelaria lacustris,* which compete with a few other species like red-coloured Cyanoprocaryota, the rhodophyte *Hildenbrandia rivularis* or certain diatoms ("Rotbunte Tiefenbiocoenose", Geitler 1928, Kann & Sauer 1982). *Bodanella lauterborni* is found only between 10 and 30 m depth on steep rocky banks in Lake Constance (Bodensee, Germany, Zimmermann 1928). *Sphacelaria lacustris* was observed at similar depths in Lake Michigan, USA (Schloesser & Blum 1980).

Freshwater brown algae tolerate a wide range of **water temperatures**. *Heribaudiella* and *Pleurocladia* from the Fischa River (Austria) were collected in a healthy growth state from January until November at temperatures from 5–13 °C. In arctic reaches at Devon Island (Canada), *Pleurocladia lacustris* endures the whole year under freezing and thawing conditions (Wilce 1966). Another locality with nearly constant temperatures of about 18–21 °C is the shaded outlet of the thermal spring of Bad Fischau (Austria); it is inhabited by *Heribaudiella*, *Pleurocladia* and the rhodophyte *Hildenbrandia* (Geitler 1932, Kusel-Fetzmann 1996). In Lake Constance, *Bodanella* grows throughout the year at 20 m depth at temperature ranges between 6 and 8 °C.
Freshwater Phaeophyta are not restricted to certain **chemical conditions;** they occur in neutral to slightly alkaline, oligotrophic to eutrophic waters. However, organic pollution is not tolerated (Geißler & Kies 2003 Knappe et al. 1996), and both *Pleurocladia* and *Heribaudiella* are classified as oligosaprobic taxa (Wegl 1983). *Heribaudiella* is widespread in streams with granite, gneiss and carbonate catchment areas. In Eastern Holstein (Germany), *Heribaudiella* and *Pleurocladia* are absent in dystrophic lime-deficient lakes surrounded by forests (2.4–6.4 mg l^{-1} Ca, pH 4.9–6.8) (Kann 1940, 1945). *Heribaudiella* is frequently found in semi-hard water lakes (15–20 mg l^{-1} Ca, pH 7.5–8.0). *Pleurocladia* was also reported for hard water lakes (>30 mg l^{-1} Ca, pH 8.0), where this taxon precipitates carbonate around the filaments (Kann 1945).
Salinity is an important factor for the occurrence of freshwater brown algae. *Pleurocladia lacustris* is nearly absent in marine systems, but widespread in freshwater habitats in northern Germany, Scandinavia, around the Baltic, Poland and Russia. Waern (1952) discovered *Pleurocladia* as an epiphyte in the Öregrund Archipelago in brackish habitats with 3.8 ‰ salinity. The thalli were comparable to freshwater specimens, but instead of a lime precipitation showed a gelatinous sheath around the filaments. *Bodanella lauterborni*, *Sphacelaria lacustris*, *S. fluviatilis* and *Heribaudiella fluviatilis* are known exclusively from freshwater lakes and rivers. If these freshwater species are descendants of marine relatives, their separation must be quite old, perhaps pre-glacial (Waern 1952).
Porterinema fluviatile was originally described from the brackish lower river courses Unter-Warnow (0.03–3.10‰) and Breitling (3.80–8.00‰) between Rostock and Warnemünde (Porter 1894, Waern 1952). Several eutrophic ponds in The Netherlands with salinity ranges from 0 to 5‰ also contained *Porterinema*. Testing with different chloride concentrations from 0 to 55‰ showed that *Porterinema* tolerates a wide euryhaline range (Dop 1979). Very recently, *Porterinema* was found in the "soda lake" Neusiedlersee (Austria; Kusel-Fetzmann & Kusel 2007), whose chemistry differs from that of brackish water (conductivity around 2000 µS cm^{-1}). The marine *Ectocarpus crouaniorum* (formerly *E. siliculosus*) is able to tolerate high chloride concentrations in waste water from potassium chloride mining in inland waters (Werra River, Germany).
The occurrence of freshwater Phaeophyceae at sites where no other algae grow luxuriantly is not caused by specific water conditions (most species are euryhaline, eurytraphent, eurytherm). Moreover, because of their very slow growth rates, phaeophyta are out-competed at sites where other algae groups dominate. This phenomenon was demonstrated by Rautenberg (1960), who conducted an experiment with test plates painted with a toxic antifouling coating. *Porterinema fluviatile* was the only colonizing species because it could tolerate the coating. On non-toxic test substrates, *Porterinema* was eventually overgrown by other fast-growing algae.

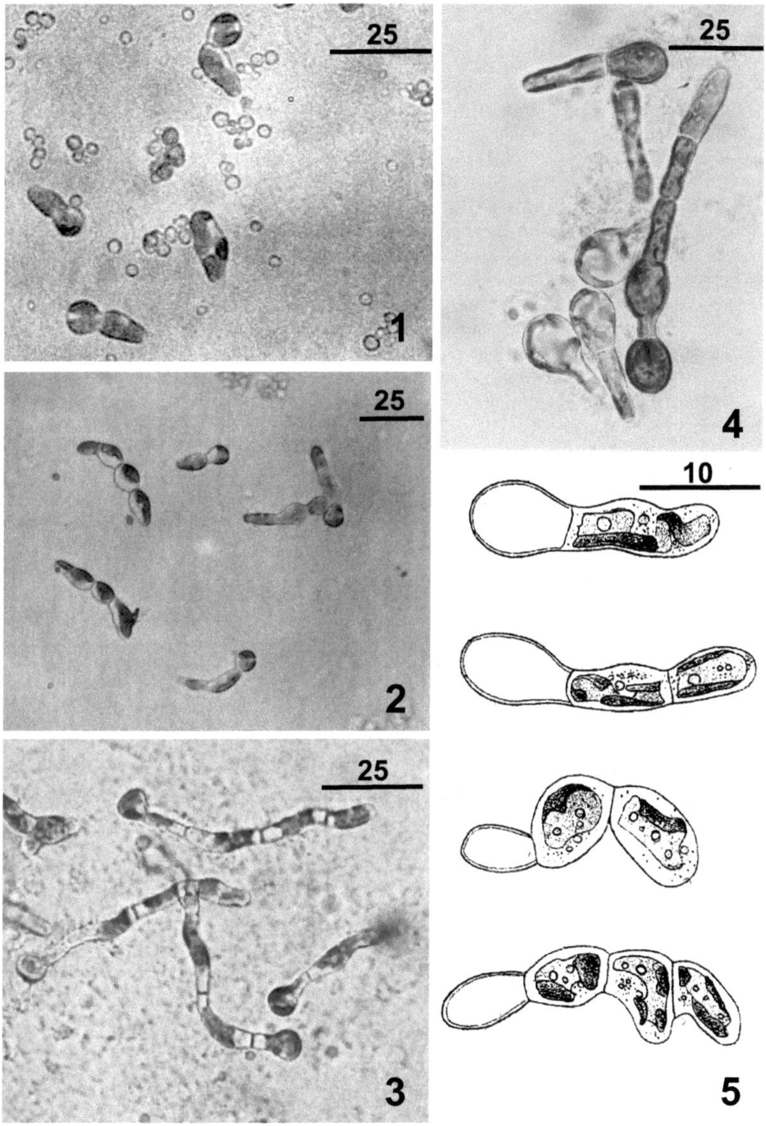

Fig. 12. Germinating swarmers settled on glass slides. **1** to **3**: *Porterinema fluviatile* (**1**, **2** Lake Neusiedlersee, Austria; **3** Channel Breitling nearby Rostock, Germany). **4**: *Ectocarpus siliculosus* from Hopkins – River (Australia). **5**: *Pleurocladia lacustris* (according to Klebahn 1895).

Sampling and Culturing

Epilithic algal crusts can be collected together with small pebbles or stone chips, or the algae material can be scraped off the substrates with a razor blade. Material should be transported in containers together with a small amount of water from the sampling site. It is recommended to collect epi- and endophytic algae together with pieces of the host plant. Additionally, a bottle of water from the site should always be taken for chemical analyses and algal cultivation.

For identification and raw cultures, samples can be kept for several weeks at around ~10 °C under dim light conditions (north-exposed window), where filaments might grow and sporangia could develop. If a glass slide is positioned between the collected raw material and the incoming light, released zoospores might settle at the glass slide and germinate.

After germlings are transferred together with the glass slide into culture medium, which should be diluted with sterile water from the sampling site, a unialgal strain may develop. Various standard media are used for culturing (Schloesser 1994, Kusel-Fetzmann & Schagerl 1992); for brackish water, 5–10% sterile sea water can be added to the media.

Glossary

Apical cell – Cell at the tip of a filament
Apical growth – Growth by division of the apical cell.
Basal growth – Division of the basal cells of free filaments or hairs.
Benthic algae – Algae that live at the bottom (= benthos) of aquatic habitats.
Biseriate – Filaments with two cell rows.
Calcified – Filaments encrusted with precipitation of calcium (and magnesium) carbonate
Conductivity – Electrical conductivity in water depends on concentration of dissolved ions and temperature. The measurement indicates approximative the ion content of the water and is usually given in Micro-Siemens per cm ($\mu S\ cm^{-1}$ – rainwater is about 5–30, typical freshwater between 100–2000, seawater above 50 000 $\mu S\ cm^{-1}$).
Diplontic life cycle – the diploid phase predominates, the haploid stage is limited to the gametes
Diplohaplontic life cycle – separate multicellular diploid (sporophyte) and haploid. (gametophyte) generations
Disjunct distribution – geographical widely separated distribution of plant or animal species.
Erect bunch – filaments arise freely and upright from the creeping. part.
Endophytic – organism living within a plant
Epiphytic – organism living on the surface of plants.
Epilithic – organism occurring on rocks and stones.
Euryhaline – living within a wide range of salinity.
Eurytherm – living within a wide temperature zone.
Eurytraphent – not bound to a certain trophic level.
Eutrophic – nutrient-rich water body.
Haplontic life cycle – the haploid phase predominates, the diploid stage is limited to the zygote.
Hard water – content of Ca^{2+} higher than 30 mg l^{-1}, semi hard water with less Ca^{2+} content.
Husk – a new sporangium or a hair, growing through the wall of a emptied sporangium; around its base the remaining sporangial wall of the older part is visible as short sheath or collar.
Intercalary growth – division of neighbouring cells along the filament.
Mastigonemata – short stiff hair like rodlets on the surface of flagellas.
Multiseriate – filaments with more than one row of cells.
Oligosaprobic – low organic matter content.
Parenchyma – tissue growing by division of cells in three directions.
Pedicell – a short, one or two cells long stalk.
Perennial – the thallus remains alive more than one year.
Plasmodesmata – thin protoplasmatic connections between adjacent cells.
Plurilocular sporangium (zoidangium) – an organ containing many small cells (loculi), from each one single spore develops (zoospore, zoid).
Propagule – a multicellular, triradiate or semi-spherical structure, a vegetative dispersal unit of *Sphacelaria*.
Prostrate – filaments growing in connection with the ground.
Pseudoparenchyma – tissue produced by closely appressed filaments.

Salinity – practical salinity units (psu) – grams of salt contained in a litre of water (A litre of seawater has around 35 g l^{-1} = 35 psu = or 35‰).
Sheath – An envelope around a filament, calcified or gelatinous, or a short basal covering with a piece of cell wall from a former dead cell.
Soda lake – a lake with sodium (and hydrogen) carbonates as predominant ions.
Supralittoral – zone above eulittoral (the zone with periodical turn of the tide) exposed to rain and heavy wave actions.
Terminal – at the end of a filament.
Uniseriate – a filament consists of only one cell row.
Unilateral – branches from filaments are one-sided.
Unilocular sporangium (or zoidium) – One special cell divides to many spores (zooids or meiospores).
Unialgal strain – only one species is cultured, but not necessarily bacteria free.

Systematic part

Class **Phaeophyceae** Kjellman 1891

Phaeophyceae De Bary 1881

Key to the genera

1a Thalli dense crusts, dark brown patches on stones, branched basal filaments with short, densely packed upright filaments and unilocular or plurilocular sporangia . . . . ***Heribaudiella***
1b Thalli not as above **2**
2a Thalli small cushions or tufts on substrata, from 0.1–20 cm in height, identifiable only with microscope **3**
2b Thalli loosely spreading with variously branched, creeping filaments **5**
3a Filaments (uni-) multiseriate . . . . ***Sphacelaria***
3b Filaments uniseriate **4**
4a Cells contain several ribbon-like plastids with pyrenoids, filaments branched, cells narrow cylindrical, spindle-like plurilocular sporangia . . . . ***Ectocarpus***
4b Short cells contain a single plastid; filaments creeping and upright, one-sided branches; unilocular ovoid-sporangia . . . . ***Pleurocladia***
5a Two plastids per cell, filaments prostrate and short erect; plurilocular sporangia crown-like, mostly consisting of four cells . . . . ***Porterinema***
5b Ten to twenty plastids per cell, filaments only prostrate, upright unilocular sporangia are present . . . . ***Bodanella***

Ectocarpus Lyngbye 1819

Long uniseriate and branched filaments arise from creeping prostrate filaments; thalli are usually 1 to 20 cm or more long. Each cell contains several ribbon-like plastids with some pear-shaped pyrenoids. The isomorphic haplo-diplontic genus *Ectocarpus* is widespread and very common along marine and brackwater coasts. More than hundred species have been described (Kützing 1843a, b, Kjellman 1890, 1891, Kylin 1947), but morphological differences of *Ectocarpus* species from sites of different salinity or with seasonal variability might have been misinterpreted as new species. Hamel (1931–1939) and Russel (1966) separated several morphotypes from the genus *Ectocarpus* and established a series of new genera like *Kuckuckia*, *Giffordia*, *Feldmannia* or *Pylaiella*. The genus *Ectocarpus* characterized by ribbon-shaped plastids and charcteristic sporangia was subdivided in two groups: the group of *E. siliculosus* and the group of *E. fasciculatus* (Hamel 1939). *E. fluviatilis* Kützing 1843 was described from a freshwater river near Monfalcone (Italy). Waern (1952) mentioned a genuine population of *E. fluviatilis*, which he found in the brackish canals of Amsterdam (Netherlands). Later on, *E. fluviatilis* was wrongly referred to as *Pylaiella littoralis forma fluviatilis* (Kützing) Hauck. *E. fluviatilis* is an uncertain species and needs to be restudied.
The remaining two groups of marine *Ectocarpus siliculosus* and *E. fasciculatus*, determined with traditional methods were confirmed by molecular studies (Stache-Crain et al. 1997). Very recently, Peters et al. (2010a, b) demonstrated that a third subclade *Ectocarpus crouaniorum* Thuret exists within the clade of *Ectocarpus siliculosus*: A strain of *E. crouaniorum* could be proved from the German river

Werra (Stache-Crain et al. 1997). *E. crouaniorum* and *E. siliculosus* cannot be differentiated with traditional methods. The most mentioned species in literature is *E. siliculosus*. Within this genus, further molecular studies are necessary to resolve the true phylogenetic relationships.

Ectocarpus siliculosus (Dillwyn 1809) Lyngb.1819 (plate **55**: fig. **4, 5**)

Ectocarpus confervoides (Roth 1797) Le Jolis 1864

Filaments irregularly branched, cells 12–40 µm broad, up to 100 µm long. Plurilocular sporangia are 12–25 µm broad and 75–200 µm long, spindle-shaped with hair-like ends. *E. siliculosus*, widespread in marine habitats, is reported in freshwaters only from the waterfalls of the Hopkins River (conductivity of 3000 µS) in Australia, 40 m above sea level and 25 km away from the marine reach (West & Kraft 1996, Entwisle et al. 1997, Mc Cauley & Wehr 2007). Sexual reproduction was never observed in freshwater; cultures from this location did not develop unilocular sporangia.
Ectocarpus confervoides, today treated as a synonym of *E. siliculosus* (Womersley 1978), was always distinguished from *E. siliculosus* in the North Sea and Baltic (Pankow 1971, Kornmann & Sahling 1977). *E. confervoides*, differing from *E. siliculosus* only in the more broad-rounded shape of the plurilocular sporangia without hair-like ending, was initially discovered in the Werra River (Germany), which showed high chloride loadings from wastes of potassium chloride mining (Geißler 1983), however, disappeared after mining ceased. The latest molecular phylogenetic analysis (Peters et al. 2010a, b) show, that the material of *E. siliculosus* from the Werra River near Hannoversch-Münden, used for the genetic study by Stache-Crain et al. (1997), was the strain (SAG: B 63.81) at the culture collection Göttingen and belong to *Ectocarpus crouaniorum* Thuret (Stache-Crain et al, 1997. To distinguish morphologically between *E. siliculosus*, *E. confervoides* and *E. crouaniorum* is hardly possible.

Pleurocladia A. Braun in Rabenhorst 1855

These small cushions of about 2 mm in diameter consist of creeping and upright filaments and can be found on stones or other substrates in streams and lakes, epiphytic on reed stems, or endophytic in macroalgae and water plants. The filaments are embedded in calcified sheaths in calcareous running waters and lakes. In brackish waters, the cushions build soft, uncalcified tufts with gelatinous sheaths around the filaments. Wilce (1966) found *Pleurocladia* at the sea coast of arctic North America, where land runoff and seawater merged intermittently, indicating that this genus is most probably euryhaline. Only one species described; a sexual life cycle is still unknown.

Pleurocladia lacustris A. Braun, in Rabenhorst 1855 (plate **56**)

Pleurocladia ralfsioides Bornet 1875 (after Flahault 1883), *Streblonema longiseta* Arnoldi 1909

The uniseriate filaments contain a single plastid (rarely two) in each cell. Near the nucleus, refractive bodies consisting probably of fucosan are abundant. Basal filaments are 16–20 µm broad; cells are nearly as long as broad. Cells of upright filaments are 8–14 (16) µm broad and 12–35 µm long. Upright filaments are often unilateral and comb-like branched (fig. **9/2**).

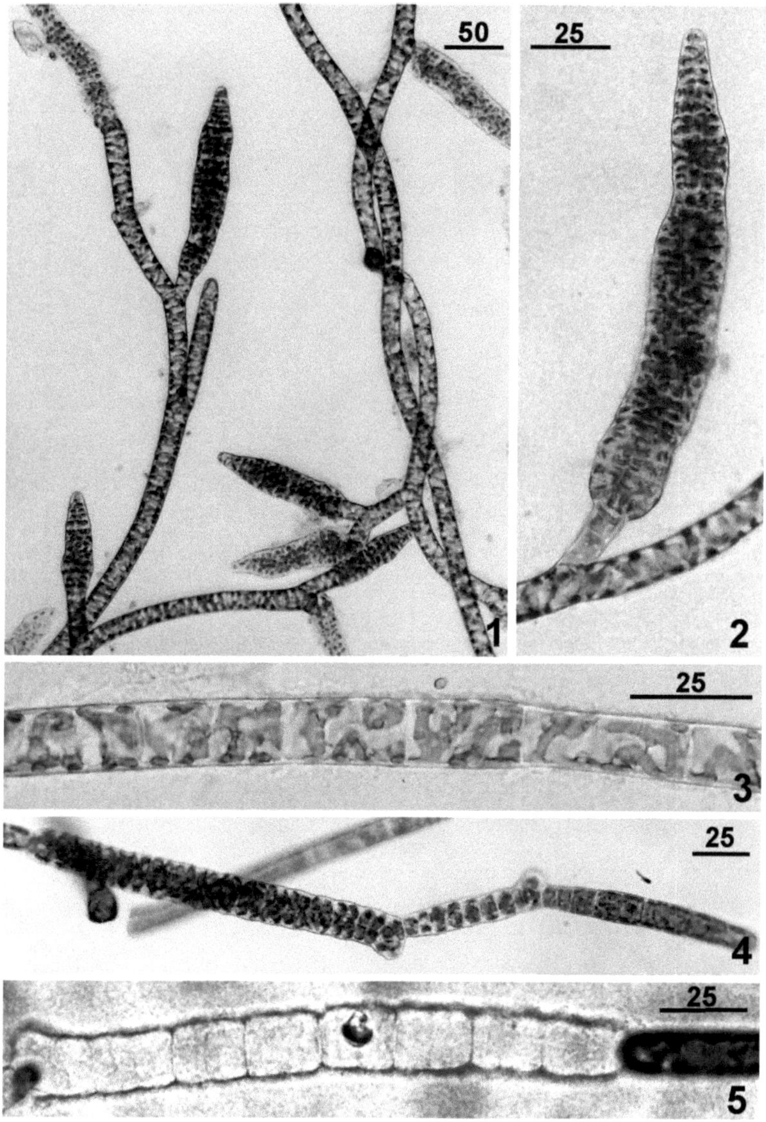

Plate 55: **Fig. 1–5.** *Ectocarpus*. **1** to **3:** *E. crouaniorum* from Werra river (Germany), **1**, **2.** Plurilocular sporangia with short stalks and rounded tips. **3.** Cells contain ribbon-like chloroplasts with pyrenoids. **4**, **5.** *E. siliculosus*, culture material from freshwater Hopkins river, Australia. **4.** Long, narrow plurilocular sporangia. **5.** Plurilocular sporangium after release of swarmers.

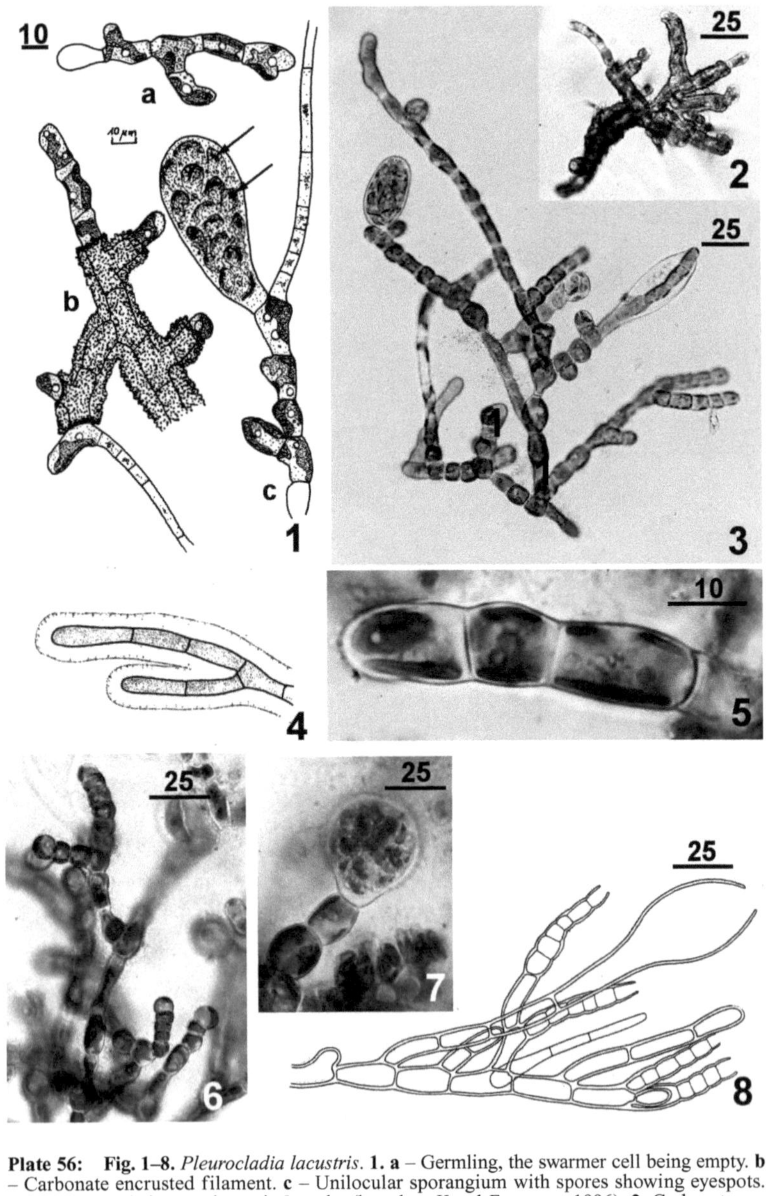

Plate 56: **Fig. 1–8.** *Pleurocladia lacustris*. **1.** **a** – Germling, the swarmer cell being empty. **b** – Carbonate encrusted filament. **c** – Unilocular sporangium with spores showing eyespots. Phaeophycean hairs are shown in **b** and **c** (based on Kusel-Fetzmann 1996). **2.** Carbonate encrusted thallus still alive. **3.** Branching part with unilocular sporangium. **4.** Branch with gelati-

Long, pale, multicellular hairs (100–300 µm), which are sometimes sheathed, arise from creeping and upright filaments; the hairs develop with a basal meristem (fig. **10/1**). Single hair cells are 5–8 µm broad and 5–10 times longer. The short side branches give rise to unilocular, globose or clavate sporangia that are 15–30 µm in diameter and 25–60–80 µm in length. The sporangia release numerous spores that eventually settle and germinate to new thalli. Uniseriate narrow plurilocular sporangia are only rarely found in freshwater habitats (Klebahn 1895, Wille 1895, Waern 1952). *Pleurocladia* was reported from lakes (Kann 1993, Szymanska & Zakrys 1990) and some rivers (Kann 1978, Kusel-Fetzmann 1996, Backhaus 2006, Backhaus & Krause 1974). *Pleurocladia lacustris* is found from freshwaters to brackish sites (Waern 1952) and in the supralittoral (Den Hartog 1968) of marine coasts. Widespread in Europe (Austria, Germany, Great Britain, Poland, Russia, Scandinavia, The Netherlands, Ukraine), USA and the Arctic (Spitzbergen, Grönland, Devon Island, Novaya Semlia). A strain can be obtained from Göttingen, SAG: B 25.93.

Sphacelaria Lyngbye 1819

Thalli consist of creeping and erect, branched, multiseriate filaments that grow by division of conspicuous apical cells. Further longitudinal cell divisions finally generate multiseriate filaments. Cells contain many small plastids without pyrenoids. *Sphacelaria* is widespread in marine habitats, where the life cycle is isomorph diplohaplontic. Two species are known from freshwater habitats; to date, they have not been reported for Europe.

Sphacelaria fluviatilis Jao 1943 (plate **57**: fig. **2**)

Small thalli up to 10 mm in height. Erect filaments with 2–6 longitudinal divisions per segment, branchings regularly opposite. Vegetative propagules are semi-spherical cell accumulations. *Sphacelaria fluviatilis* was initially described from Kialing River in Western China (Jao 1943, 1944), growing submerged on rocks on shallow banks of the river, which were air-exposed at low water levels. The species was later found in Gull Lake (Michigan, USA; Thompson 1975). Timpano (1978, 1980) studied its life cycle and ultrastructure. Neither plurilocular nor unilocular sporangia could be observed. Contrary to the description of Jao, the American populations showed hairs.

Sphacelaria lacustris Schloesser & Blum 1980 (plate **57**: fig. **1**)

The small semi-spherical cushions are under 2 mm and appear as calcified thalli. Schloesser & Blum (1980) discovered this species on rocks in the sublittoral (5–15 m depth) of the western shore of Lake Michigan, USA, which is still the only finding. The growth of filaments occurs through division of apical cells, which are 12–13 µm broad and 30–60 µm long. The plastids are located in the tip, which also contains numerous physodes (small, highly refractive fucosan-vesicles). From

---

nous sheath, material from brackish water (based on Klebahn 1895). **5.** Cells with one chloroplast each. **6.** Short branches with four cells could be plurilocular sporangia. **7.** Rounded unilocular sporangium. **8.** One-sided branches, empty unilocular and plurilocular (short row of cells) sporangia (based on Klebahn 1895).

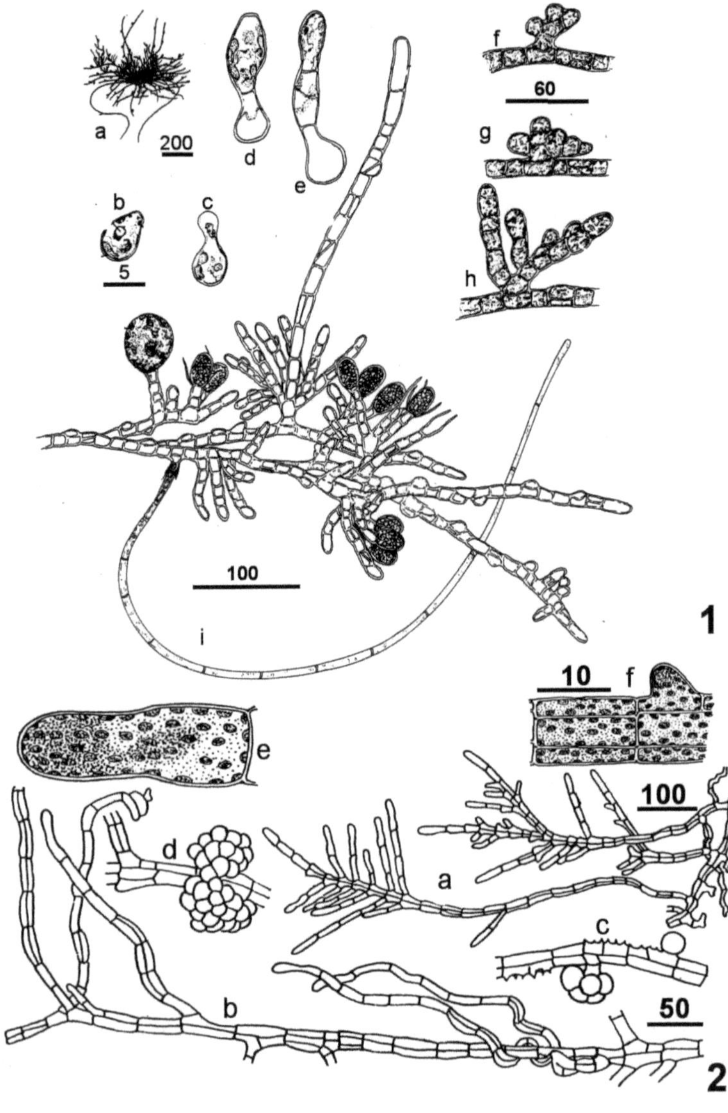

Plate 57: **Fig. 1.** *Sphacelaria lacustris*. **a** – Typical appearance of thallus, b – spore, **c**, **d**, **e** – spore germination, **f**, **g**, **h** – stages of propagule formation, **i** – basal part with unilocular sporangia, an upright filament, which is partly biseriate showing an apical cell, and a sheated hair with basal meristem (based on Schloesser & Blum 1980). **Fig. 2.** *Sphacelaria fluviatilis*. **a** – Thallus exposed to air at low water level, **b** – multiseriate thallus with rhizoids, growing below water at high water level, **c**, **d** – propagules, **e** – apical cell, **f** – part of a multiseriate filament (based on Jao 1943).

prostrate, creeping filaments and erect filaments with a maximum length of about 1 mm, multicellular sheated hairs develop; they have a basal meristem. The filaments are commonly uniseriate, sometimes also in parts biseriate. In contrast to *S. fluviatilis,* irregular branches occur sporadically, but are never opposite. Vegetative reproduction involves triradiate propagules. Unilocular sporangia (60 µm in diameter) released non-motile spores in culture, but fusion was not observed (Schloesser & Blum 1980).

Heribaudiella Gomont 1896

Lithoderma Areschoug 1875

Heribaudiella is the most frequently mentioned freshwater Phaeophyceae and can be recognized with the naked eye. The crust-forming alga shows brownish patches on stones and rocky areas in clear springs, fast-flowing brooks, streams and lakes. *Heribaudiella* crusts are mentioned from many freshwater streams (also from mountain areas) far away from the sea. *Heribaudiella* prefers rocks (granite, gneiss, basalt, limestone; Wehr & Stein 1985, Wehr 2003), but it also colonizes artificial substrata like bricks (Svedelius 1926, 1930, Kusel-Fetzmann 1996). One species is described; sexual reproduction is unknown.

Heribaudiella fluviatilis (Areschoug) Svedelius 1930 (plate **58**)

Ralfsia fluviatilis Areschoug 1871, *Lithoderma fluviatile* Areschoug 1875, *Lithoderma fontanum* Flahault 1883, *Heribaudiella arvernensis* Gomont 1896, *Lithoderma zonatum* Jao 1941

The first descriptions stem from collections taken at various sites with different ecology (Scandinavia, France). Because the organisms differed in the number of cells in vertical cell rows, some new taxa were described initially. Further studies, however, confirmed that these descriptions merely represented morphotypes of a single species. (Hansgirg 1905, Skuja 1925, Waern 1938, 1949, Israelson 1938). The brownish thalli consist of creeping and repeatedly branching filaments that form closed discs; cells are 6–10 µm broad and 12–32 µm long. The discs give rise to vertical filaments, which are densely packed and 5 to 15 (20) cells high. These cells are 8–15 µm in diameter. Each cell contains 6–10 plastids. At the tip of vertical filaments, unilocular sporangia may develop (width 12–15 µm, length 24–28 µm). After spore release, a hair or a new sporangium can grow through the remaining cell wall (husk), which serves as a basal sheath. Settled spores develop into creeping filaments, soon forming small crusts with a few hairs (fig. **10/5**). Plurilocular sporangia are also produced at the tip of a cell row. The species is perennial, and a strain is available at Göttingen SAG B 13.90.

Bodanella Zimmermann 1928

Thalli are small, brownish cushions of creeping threads on steep limestone rocks in deeper lake areas (10–30 m). Besides the Bodensee (Lake Constance, Germany), where this genus was initially described by Zimmermann (1928), the organism was found only in some Austrian pre-alpine lakes (Lunzer See, Traunsee and Zellersee; Geitler 1928, Kann 1982 a, b). *Bodanella* has probably lost its ability to reproduce sexually (Müller & Geller 1978). Only one species is described.

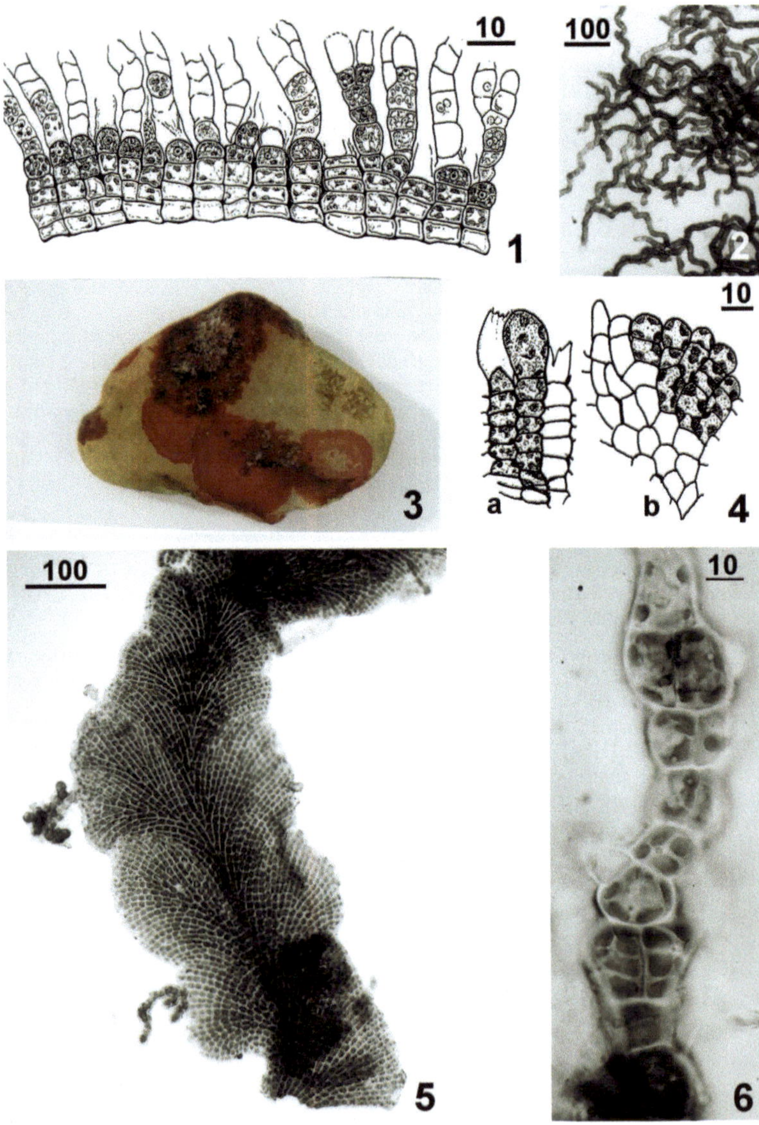

Plate 58: **Fig. 1–6.** *Heribaudiella fluviatilis*. **1.** Cross section of a crust with rows of emptied plurilocular sporangia (based on Svedelius 1930). **2.** Filaments, not fixed at the glass in culture are freely growing in the water. **3.** Crusts of *Heribaudiella* (top, dark) and the rhodophyte *Hildenbrandia* (below, smaller, rounded crusts, red) **4. a** – Cross section of a crust with uni-

Bodanella lauterborni Zimmermann 1928 (plate 59)

The uniseriate filaments with apical growth develop, in contact with the substrate, irregularly pseudoparenchymatous discs, but without erect filaments. The cells are short, sometimes inflated, rounded or bent (10–16 µm wide and 10–25 µm long) and contain 10–15 discoid parietal plastids. Unilocular, globose to ovoid sporangia develop mostly directly from the filaments or with 1 or 2 cells as a stalk system (pedicells).
Thalli from natural sites never developed hairs, but in culture thin hairs (6–10 µm in diameter) with basal growth (fig. **10/3**) were often observed (Müller & Geller 1978). Under culture conditions, about 16 slow-moving swarmers are released from unilocular sporangia. After settling, new thalli with unilocular sporangia grow. Plurilocular sporangia were never observed (Müller & Geller 1978). *Bodanella lauterborni* can be obtained from Göttingen SAG: B 123.79.

Porterinema Waern 1952

Porter (1894) discovered this organism in the slightly brackish Unter-Warnow River (Germany) near its mouth into the Baltic and initially described it as *Streblonema fluviatile*. The thalli consist of small, irregularly branched filaments and are endophytic (in *Enteromorpha* and *Potamogeton pectinatus*), epiphytic (on *Phragmites*), epizootic (on hydroids) or epilithic. Various cells from creeping filaments produce upright branches with sporangia or hairs. Usually, plurilocular sporangia show a crown-like arrangement with 4 cells (sometimes more); they are intercalar, sessile or develop at one- or two-celled pedicells.
Based on the unique sporangia, differing from the genus *Streblonema*, Waern (1952) described this genus as *Porterinema*. It is widespread in waters with elevated salinity, but has never been found in pure freshwaters or typical marine environments (Wilce et al. 1970). Therefore this species can be regarded as an exclusively brackish species (Waern 1952).

Porterinema fluviatile (Porter) Waern 1952 (plate **60, 61**)

Streblonema fluviatile Porter 1894, *Porterinema marina* Jaasund 1965, *Pseudobodanella peterfii* Gerloff 1967, *Apistonema pyrenigerum* Pascher 1931, *Apistonema expansum* Geitler 1970

Creeping thalli form loose discs of radiating branched filaments (fig. **9/1**) and short erect filaments. The inflated cells are usually 6–12 µm in diameter and of similar length. Sometimes, cells of erect filaments can be up to 50 µm long. Each cell contains two, sometimes three plate-like plastids with pyrenoids (visible only using a transmission electron microscope, fig. **8**). The multi-celled hairs with basal growth are 3–8 µm wide, up to 200 µm long and sheathed at their base (fig. **10/2**). The pear-shaped unilocular sporangia occur only sporadically (width 15–30 µm; length 80 µm); plurilocular sporangia are abundant. They may develop in filaments intercalary, sessile, on one or two pedicells, or terminal. After repeated division, groups of up to 32 sporangia can be found. New thalli develop from zoospores, and sexual reproduction has never been observed.

---

locular sporangia, some of them already empty. **b** – Edge of a young crust (based on Kusel-Fetzmann 1996). **5.** Pseudoparenchymatous, fan-like crust on a glass surface (culture material). **6.** Young filament starting to develop multicellular rows.

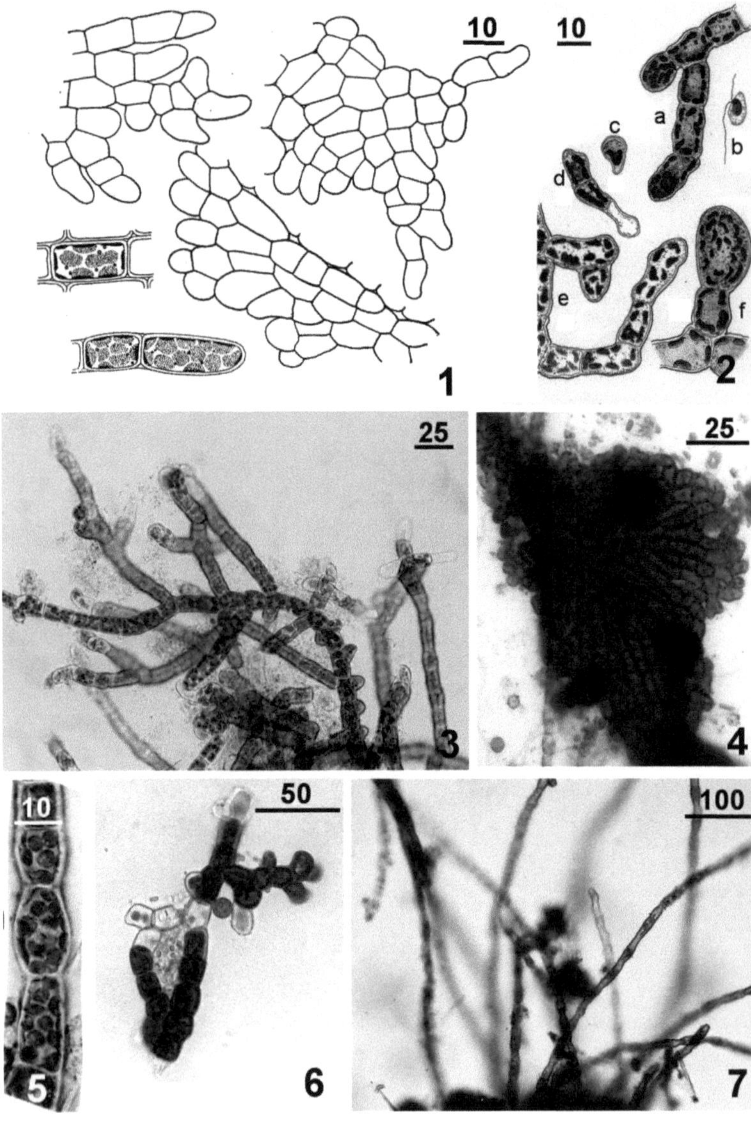

Plate 59: **Fig. 1–7.** *Bodanella lauterborni*. **1.** Pseudoparenchymatous disc with apical growth, and details of cells (based on Geitler 1928). **2. b** – Drawings of a swarmer, **c**, **d** – germlings, **a**, **e** – short upright vegetative filaments, **f** – branch with unilocular sporangium (based on the original description from Zimmermann 1928). **3.** Loosely creeping and short upright filaments. **4.** Pseudoparenchymatous layer attached to the bottom. **5.** Living cells with chloroplasts. **6.** Short rows of cells, possibly developing unilocular sporangia. **7.** Long filaments growing from the bottom layer into the surrounding water.

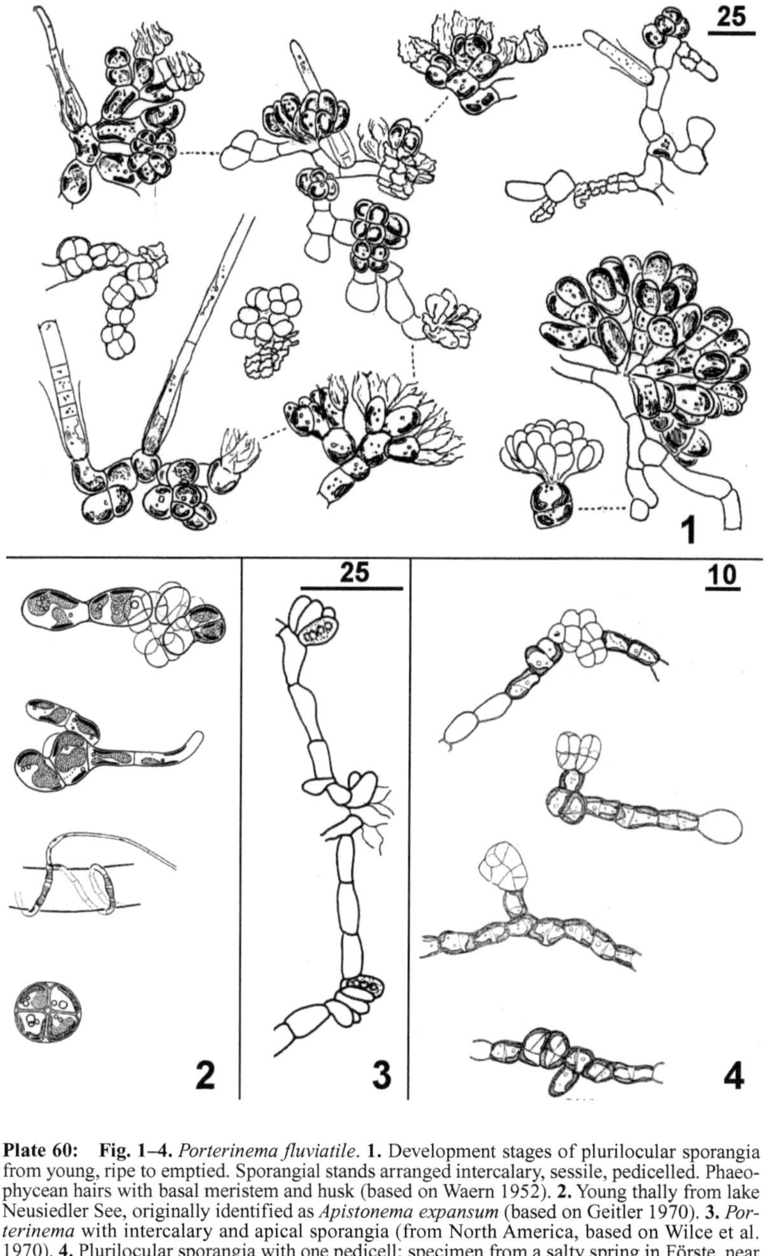

Plate 60: Fig. 1–4. *Porterinema fluviatile*. **1.** Development stages of plurilocular sporangia from young, ripe to emptied. Sporangial stands arranged intercalary, sessile, pedicelled. Phaeophycean hairs with basal meristem and husk (based on Waern 1952). **2.** Young thally from lake Neusiedler See, originally identified as *Apistonema expansum* (based on Geitler 1970). **3.** *Porterinema* with intercalary and apical sporangia (from North America, based on Wilce et al. 1970). **4.** Plurilocular sporangia with one pedicell; specimen from a salty spring in Förste, near Harz Mountains (based on Gerloff 1967).

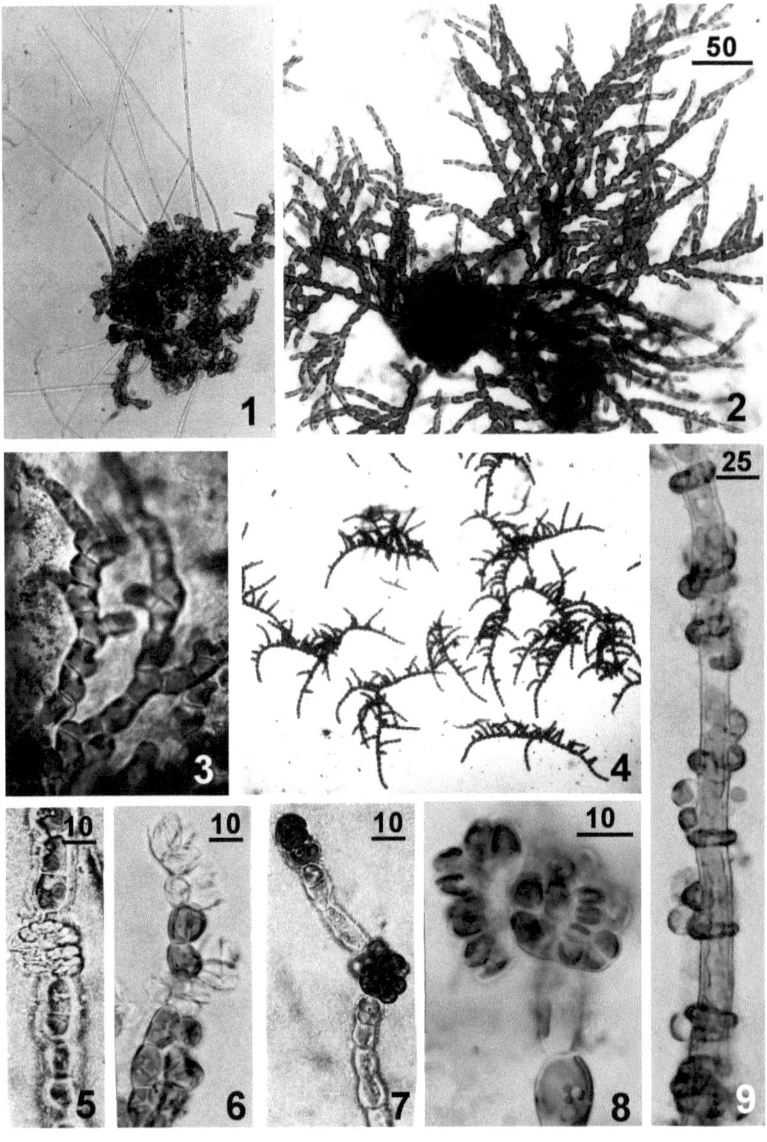

Plate 61: **Fig. 1–9.** *Porterinema fluviatile*, living material. **1.** Thallus produced many phaeophycean hairs as a consequence of nutrient storage. **2.** Basal layer with radiate creeping filaments, centre elevated. **3.** Epizootic filaments creeping on a hydrocaule of *Cordylophora caspia* (Hydrozoa). **4.** Young thalli in culture after germination of spores and settling at glass slide. **5, 6.** Empty plurilocular sporangia, intercalary and apical. **7.** Ripe plurilocular sporangia. **8.** Group of plurilocular sporangia. **9.** A young *Porterinema* filament creeps around a *Cladophora* filament.

Porterinema fluviatile was reported from The Netherlands in a peat pond with 0.4‰ salinity (Dop & Vroman 1976). Sites further inland that are inhabited by *Porterinema* are saltwater springs near Artern (<5‰, Germany, Rieth 1969) and the Werra River at Bad Sooden (Germany, Schlösser 1994). Gerloff (1967) described *Pseudobodanella peterfii* from a saltwater spring with high calcium sulphate and carbonate concentration in Förste near Osterode (Germany). According to Bourrelly (1981), this species is synonymous with *Porterinema fluviatile*. Very recently, *Porterinema* was found in the soda lake Neusiedlersee (Austria), creeping on the hydrocaule of the brackish hydrozoan *Cordylophora caspia* (Kusel-Fetzmann & Kusel 2007). For lake Neusiedlersee, this alga was described by Geitler (1970) as the chrysophyte *Apistonema expansum*, growing epiphytically on reed roots. Dop (1979) identified this and other Chrysophyceae *Apistonema* (Pascher 1931) as synonymous to the Phaeophyceae *Porterinema*. Strain available at Göttingen SAG: B 124. 79

References to Phaeopyceae

Arnoldi W. (1909): *Streblonema longiseta* n. spec. – Flora **99**: 465–472.

Areschoug J.E. (1875): Observationes Phycologicae, III. Acta Reg. Soc. Sci, Upsala, Ser. **III**, 10, 1–35, Plates 1–3.

Backhaus D. (2006): Litorale Aufwuchsalgen im Hoch- und Oberrhein. – Carolinea **64**: 5–68.

Backhaus D. & Krause W. (1974): Beiträge zu einer Algenflora des Taubergießengebietes. – Taubergießengebiet, die Natur- und Landschaftsschutzgebiete Baden-Württembergs, **7**: 177–179.

Borisova O.V. (2006): Phaeophyta. In: Algae of Ukraine: Diversity, Nomenclature, Taxonomy, Ecology and Geography. Edited by Tsarenko P.M., Wasser S.P. & Nevo E., Ruggell A.R. Gantner Verlag K.G.

Bourrelly P. (1981): Les Algues d'Eau Douce. Tom II. Les Algues jaunes et brunes Chrysophycées, Phéophycées, Xanthophycées et Diatomées. Rev. Edn., Soc. Nouvelle des Éditions Boubée, Paris.

Braun A. (1854): *Pleurocladia*. – In: Rabenhorst, L. (1855): Die Algen Sachsens Nr. **441**.

Budde H. (1927): Die Rot- und Braunalgen des Westfälischen Sauerlandes. – Ber. Deutsch. Bot. Ges. **45**: 143–150.

Den Hartog C. (1968): The littoral environment of rocky shores as a border between the sea and the land and between the sea and the fresh water. – Blumea **16**(2): 375–393.

Dillwyn L.W. (1802–1809): British Confervae or colored figures and descriptions of the british plants. London.

Dop A.J. & Vroman M. (1976): Observations on some interesting freshwater algae from the Netherlands. – Acta Bot. Neerlandica **25**(5): 321–328.

Dop A.J. (1979): *Porterinema fluviatile* (Porter) Waern (Phaeophyceae) in the Netherlands. – Acta Bot. Neerlandica **28**(6): 449–458.

Draisma, S.G.A., Prud'homme van Reine W.F., Stam W.T. & Olsen J.L. (2001): A Reassessment of phylogenetic Relationsships within the phaeophyceae based on Rubisco large subunit and ribosomal DNA sequences. – J. Phycol. **37**: 586–603.

Entwisle T.J., Sonneman J.A. & Lewis S.H. (1997): Freshwater Algae in Australia. A Guide to conspicuous Genera. Sainty and Associates Pty Ltd, Potts Point, NSW, Australia.

Flahault M.C. (1883): Sur le *Lithoderma fontanum*, algue Phaéeosporée d'eau douce. – Bull. Soc. Bot France **30**: 102–107.
Fritsch F.E. (1929):The encrusting algal communities of certain fast flowing streams. – New Phytologist **28**: 165–196.
Fritsch F.E. (1965): The structure and Reproduction of the Algae. Vol. II Phaeophyceae, Rhodophyceae, Myxophyceae. Cambridge Univ. Press, Cambridge, pp. 939.
Geißler U. (1983): Die salzbelastete Flußstrecke der Werra – ein Binnenlandstandort für *Ectocarpus confervoides* (Roth) Kjellman. – Nova Hedwigia **37**: 193–217.
Geißler U. & Kies L. (2003): Artendiversität und Veränderungen in der Algenflora zweier städtischer Ballungsgebiete Deutschlands: Berlin und Hamburg. – Nova Hedwigia Beiheft **126**, 777 p.
Geitler L. (1928): Über die Tiefenflora an Felsen im Lunzer Untersee. – Arch. Protistenkunde **62**: 96–104.
Geitler L. (1932): Notizen über *Hildenbrandia rivularis* und *Heribaudiella fluviatilis*. – Arch. Protistenkunde **76**: 581–588.
Geitler L. (1970): Beiträge zur epiphytischen Algenflora des Neusiedler Sees. – Österr. Bot. Z. **118**: 17–29.
Gerloff J. (1967): Eine neue Phaeophyceae aus dem Süßwasser: *Pseudobodanella peterfii* nov. gen. et nov. spec. – Rev. Roum., Biol. Sér. Bot. **12**(1): 27–35.
Gomont M. (1896): Contribution à la Flore Algologique de la Haute-Auvergne. – Bull. Soc. Bot. France **43**, Ser. 3: 373–393.
Graham L.E. & Wilcox L.W. (2000): Algae. Prentice Hall, Inc.
Graham L.E., Graham J.M. & Wilcox L.W. (2009): Algae. Second Edition, Benjamin Cummings Pearson.
Hamel G. (1931–1939a): Phaeophycées de France. Trait de Botanique **47**, Paris, 432 pp.
Hamel G. (1939b): Sur la classification des Ectocarpales. – Bot. Notiser **1939**: 65–70.
Hansgirg J.P.C. (1905): Grundzüge der Algenflora von Niederösterreich. – Beih. Bot. Zentralblatt **18**: 417–522.
Hauck F. (1885): Die Meeresalgen Deutschlands und Österreichs. – In: Rabenhorst L. Die Kryptogamenflora von Deutschland, Österreich und der Schweiz, Band **2**. Leipzig.
Hoppe H.A. (1962): Geschichtlich Entwicklung der Auswertung von Meeresalgen. Botanica Marina, vol **III** Suppl.
Israelson G. (1938): Über die Süßwasserphaeophyceen Schwedens. – Bot. Notiser **1938**: 113–128.
Jao C.-C. (1941): Studies on the freshwater algae of China. VII. *Lithoderma zonatum*, a new freshwater member of the Phaeophyceae. – Sinensia **12**: 239–244.
Jao C.-C. (1943): Studies on the freshwater algae of China. XI. *Sphacelaria fluviatilis*, a new freshwater brown alga. – Sinensia **14**: 151–154.
Jao C.-C. (1944): Studies on the freshwater algae of China. XII. The attached algal communities of the Kialing River. – Sinensia **15**: 61–73.
Jaasund E. (1965): Aspects of the marine algal vegetation of North Norway. – Botanica gothoburg. **4**: 1–174.
Kann E. (1940): Ökologische Untersuchungen an Litoralalgen ostholsteinischer Seen. – Arch. Hydrobiol. **37**: 177–269.
Kann E. (1945): Zur Ökologie der Litoralalgen in ostholsteinischen Waldseen. – Arch. Hydrobiol. **41**: 14–42.

Kann E. (1978): Systematik und Ökologie der Algen Österreichischer Bergbäche. – Arch. Hydrobiol. Suppl. **53**: 405–643.
Kann E. (1982a): Qualitative Veränderungen der litoralen Algenbiocönose österreichischer Seen (Lunzer Untersee, Traunsee, Attersee) im Laufe der letzten Jahrzehnte. – Arch. Hydrobiol. Suppl. **62** (Monographische Beiträge): 440–490.
Kann E. (1982b): Die litorale Algenvegetation des Zeller Sees (Salzburg, Österreich). – Arch. Hydrobiol. **94**: 492–501.
Kann E. (1993): Der litorale Algenaufwuchs im See Erken und in seinem Abfluss (Upland, Schweden). – Algolog. Studies **69**: 91–112.
Kann E. & Sauer F. (1982): Die „Rotbunte Tiefenbiocoenose" (Neue Beobachtungen in österreichischen Seen und eine zusammenfassende Darstellung). Arch. Hydrobiol. **95**: 181–195.
Kann E. & Tschamler H. (1976): Algenaufwuchs unter natürlichen Bedingungen auf Kunststoffen. – Chemie/Kunststoffe aktuell **2/1976**: 63–71.
Kirkby S.M., Hibberd D.J. & Whitton B.A. (1972): *Pleurocladia lacustris* A. Braun (Phaeophyta) – a new British Record. – Vasculum **57**: 51–56.
Kjellman F.R. (1890): Handbok i Skandinaviens hafsalgflora I Fucoideae Stockholm.
Kjellman F.R. (1891): Choristocarpaceae. – In: Engler A. & Prantl K. (eds): Die natürlichen Pflanzenfamilien 1. Teil, 2. Abt., p. 191, Leipzig.
Klebahn H. (1895): Beobachtungen über *Pleurocladia lacustris* A. Br. – Ber. Deutsch. Bot. Ges. **13**: 93–106.
Knappe J., Geissler U., Gutowsky A. & Friedrich G. (1996): Rote Liste der limnischen Braunalgen (Fucophyceae) und Rotalgen (Rhodophyceae) Deutschlands. – Schr.-R. f. Vegetationskunde H. **28**: 609–623.
Kornmann P. & Sahling P.H. (1977): Meeresalgen von Helgoland. Benthische Grün-, Braun- und Rotalgen. Phaeophyceae. – Helgoländer wissenschaftliche Meeresuntersuchungen **29**: 1–289.
Kusel-Fetzmann E.L. (1996): New Records of freshwater Phaeophyceae from lower Austria. – Nova Hedwigia **62**: 79–89.
Kusel-Fetzmann E.L. & Schagerl M. (1992): Verzeichnis der Sammlung von Algenkulturen an der Abteilung für Hydrobotanik am Institut für Pflanzenphysiologie der Universität Wien. – Phyton **32**: 209–234.
Kusel-Fetzmann E.L. & Kusel H. (2007): *Cordylophora caspia* Pallas – Erstfund im Neusiedlersee. – Verhandl. Zool.-Bot. Ges. Österr. **144**: 39–47.
Kützing F.T. (1843a): Tabulae Phycol., Vol. **5**, Genus Ectocarpus.
Kützing F.T. (1843b): Phycologia generalis, Ectocarpeae, Fam. XXV Ectocarpaceae 287–290. Leipzig, F.A. Brockhaus.
Kylin H. (1947): Die Phaeophyceen der schwedischen Westküste. Lunds Univ. Årsskrift N.F. Avd. **2, 43** Lund 1–99.
Le Jolis A. (1864): Liste des algues marines de Cherbourg. – Mém. Imp. Soc. Sci. Nat. Cherbourg **10**: 1–168.
Lee R.E. (1999): Phycology, 3. Edition Cambridge Univ. Press, Cambridge.
Lüning K. (1985): Meeresbotanik. Verbreitung, Ökophysiologie und Nutzung der marinen Makroalgen. Georg Thieme Verlag, Stuttgart – New York, 375 p.
Lyngbye H.C. (1819): Tentamen Hydrophytologiae Danicae. Hafniae.
McCauley L.A.R. & Wehr J.D. (2007): Taxonomic reappraisal of the freshwater brown algae *Bodanella, Ectocarpus, Heribaudiella and Pleurocladia* on the basis of rbcL sequences and morphological characters. – Phycologia **46**(4): 429–439.

Müller D.G. &. Geller W. (1978): Einige Beobachtungen an Kulturen der Süßwasser-Braunalge *Bodanella lauterborni* Zimmermann. – Nova Hedwigia **29**: 735–741.
Oberdorfer E. (1928): Lichtverhältnisse und Algenbesiedlung im Bodensee. – Zeitschr. f. Bot. **20**: 465–567.
Pankow H. (1971): Algenflora der Ostsee, I. Benthos. Gustav Fischer Verlag, Stuttgart.
Pascher A. (1925): Phaeophyta (Phaeophyceae). – In: Die Süßwasserflora Deutschlands, Österreichs und der Schweiz, Heft **11**: 119–133.
Pascher A. (1931): Eine neue braune Fadenalge des Süßwassers. – Archiv f. Protistenk. **73**: 60–71.
Peters A.F., Van Wijk S.J., Cho G.Y., Scornet D., Hanyuda T., Kawai H., Schroeder D.C., Cock J.M. & Boo S.M. (2010a): Reinstatement of *Ectocarpus crouaniorum* Thuret in Le Jolis as a third common species of *Ectocarpus* (Ectocarpales, Phaeophyceae) in Western Europe, and its phenology at Roscoff, Brittany. – Phycological Research **58**: 157–170.
Peters A.F., Mann A.D., Córdova C.A., Brodie J., Correa J.A., Schroeder D.C. & Cock J.M. (2010b): Genetic diversity of *Ectocarpus* (Ectocarpales, Phaeophyceae) in Peru and northern Chile, the area of origin of the genome-sequenced strain. New Phytologist. doi: 10.1111/j. 1469-8137.2010.03303x.
Porter H.Ch. (1894): Abhängigkeit der Breitling- und Unterwarnow-Flora vom Wechsel des Salzgehaltes. Diss. Rostock.
Pueschel C.M. & Stein J.R. (1983): Ultrastructure of a Freshwater Brown Alga from Western Canada. – J. Phycol. **19**: 209–215.
Rabenhorst L. (1855): Algen Sachsens exs. dec. **45**, Nr. **441**.
Rautenberg E. (1960): Zur Morphologie und Ökologie einiger epiphytischer und epi-endophytischer Algen. – Botanica Marina **II**: 133–148.
Reviers B. de, Rousseau F. & Draisma S.G.A. (2007): Classification of the phaeophyceae from past to present and current challenges. – In: Brodie J. & Lewis J. (eds.), Unravelling the algae, the past, present and future of algal systematics. By Systematics Association Special Vol. Series **75**: 267–284.
Riet A. (1969): Über eine auf Vaucherien im Arterner Salzgebiet epiphytische Alge und die durch sie aufgeworfene Problematik. – Arch. Protistenkunde **111**: 129–137.
Roth A.G. (1797): Catalecta botanica. Fasc. **I**.
Russell G. (1966): The Genus *Ectocarpus* in Britain, I. Attached forms. – J. mar. biol. Ass. U.K. **46**: 267–294.
Schlösser U.G. (1994): SAG-Sammlung von Algenkulturen at the University of Göttingen, Catalogue of strains 1994. – Botanica Acta **107**: 113–186.
Schloesser R.E. & Blum J.L. (1980): *Sphacelaria lacustris* spec. nov. A Freshwater Brown Alga from Lake Michigan. – J. Phycol. **16**: 201–207.
Schmid O.J. & Hoppe H.A. (1962): Meeresalgen als Dünge-, Futter- und Nahrungsmittel. In: Meeresalgen. Industrielle Bedeutung und Verwendung. – Botanica Marina **III**, Suppl. 101–119.
Silva P.C. & De Reviers B. (2000): Ordinal names in the Phaeophyceae. – Cryptogam. Algol. **21**: 49–58.
Skuja H. (1925): Bemerkungen über die Süßwasserarten der Gattung *Lithoderma* in Lettland IV – Hedwigia **65**: 331–340.
Stache-Crain B., Müller D.G. & Goff L.J. (1997): Molecular systematics of *Ectocarpus* and *Kuckuckia* (Ectocarpales, Phaeophyceae) inferred from phylogenetic analysis of nuclear- and plastid-encoded DNA sequences. – J. Phycol. **33**: 152–168.

Starmach K. (1977): Phaeophyta, Rhodophyta. In: Starmach K. (ed.), Flora slodkowodna Polski T 14 PWN Warszawa & Krakow, 445 p.
Svedelius N. (1926): *Lithoderma fluviatile* i Sverige. – Svensk Bot. Tidskrift **20**: 494.
Svedelius N. (1930): Über die sogenannten Süßwasser Lithodermen. – Z. Bot. **23**: 892–918.
Symoens J.J. (1950): Esquisse d'un systeme des associations algales d'eau douce. – Int. Ver. Theor. Angew. Limnol. Verh. **11**: 395–408.
Szymanska H. & Zakrys B. (1990): New phycological records from Poland. – Algological Studies **60**: 25–32.
Timpano P. (1978): A preliminary report of the freshwater phaeophyte *Sphacelaria fluviatilis*. – J. Phycol. **14**(2), Suppl. Abstract **86**, p. 34.
Timpano J. (1980): The ultrastructure of the freshwater phaeophyte *Sphacelaria fluviatilis*. – J. Phycol. **16**, Suppl., Abstract **167**, p. 44.
Thompson R.H. (1975): The freshwater Brown Alga *Sphacelaria fluviatilis*. – J. Phycol. **11**, Suppl., p. 5.
Van den Hoek C., Mann D.G. & Jahns H.M. (1995): Algae. An introduction to phycology. Cambridge Univ. Press, Cambridge.
Waern M. (1938): Om *Cladophora aegagropila*, *Nostoc pruniforme* och andra alger i Lilla Ullevifjärden, Mälaren. – Bot. Notiser **1938**: 129–142.
Waern M. (1949): Remarks on Swedish *Lithoderma*. – Svensk. Bot. Tidskr. **43**: 633–670.
Waern M. (1952): Rocky-Shore Algae in the Öregrund Archipelago. – Acta Phytogeographica Suecica **30**: 94–173.
Wegl R. (1983): Index für die Limnosaprobität. – Wasser und Abwasser **26**: 1–175.
Wehr J.D. (2003): Brown Algae. – In: Wehr J.D. & Sheath R.G. (eds.), Freshwater Algae of North America, 757–773.
Wehr J.D. & Perrone A.A. (2003): A new record of *Heribaudiella fluviatilis*, a freshwater brown alga (Phaeophyceae), from Oregon. – Western North American Naturalist **63**(4): 517–523.
Wehr J.D. & Stein J.R. (1985): Studies on the Biogeography and Ecology of the freshwater Phaeophycean Alga *Heribaudiella fluviatilis*. – J. Phycol. **21**: 81–93.
West J.A. & Kraft G.T. (1996): *Ectocarpus siliculosus* (Dillwyn) Lyngbye from Hopkins River Falls, Victoria – the first record of a freshwater brown alga in Australia. – Muelleria **9**: 29–33.
Wilce R.T. (1966): *Pleurocladia lacustris* in Arctic America. J. Phycol. **2**: 57–66.
Wilce R.T., Webber E.E. & Sears J.R. (1970): *Petroderma* and *Porterinema* in the New World. – Marine Biology **5**: 119–135.
Wille N. (1895): Über *Pleurocladia lacustris* A. Braun und deren systematischer Stellung. – Ber. Deutsch. Bot. Ges. **8**: 106–112.
Womersley H.B.S. (1987): The marine benthic Flora of Southern Australia. Part II, Phaeophyta and Genus *Vaucheria* (Chrysophyta). South Australian Government Printing Division, Adelaide, 484 p.
Zimmermann W. (1928): Über Algenbestände aus der Tiefenzone des Bodensees. Zur Ökologie und Soziologie der Tiefseepflanzen. – Z. Bot. **20**: 1–28.

Index

In the following list, currently valid names are in **bold**, names of genera and lower taxa are in ***bold italics***. Taxa in normal italics are treated as old names or synonyms. Page numbers in **bold** refer to the main discussion of the taxon.

MIX
Papier aus verantwortungsvollen Quellen
Paper from responsible sources
FSC® C105338

If you have any concerns about our products,
you can contact us on
ProductSafety@springernature.com

In case Publisher is established outside the EU,
the EU authorized representative is:
Springer Nature Customer Service Center GmbH
Europaplatz 3, 69115 Heidelberg, Germany

Printed by Libri Plureos GmbH
in Hamburg, Germany